PERGAMON INTERNATIONAL LIBRARY
of Science, Technology, Engineering and Social Studies
The 1000-volume original paperback library in aid of education, industrial training and the enjoyment of leisure
Publisher: Robert Maxwell, M.C.

ROCK BOLTING

A Practical Handbook

Other Titles of Interest

JANSSON, B. & WINQVIST, T.
Planning of Subsurface Use

KITAMURA, I.
Tunnelling Under Difficult Conditions

SMITH, R. A.
Fracture Mechanics

Related Journals

Underground Space

International Journal of Rock Mechanics and Mining Sciences and Geomechanics

ROCK BOLTING

A Practical Handbook

Edited by

R. SCHACH

K. GARSHOL

and

A. M. HELTZEN

Norwegian Institute of Rock Blasting Techniques

PERGAMON PRESS

OXFORD · NEW YORK · TORONTO · SYDNEY · PARIS · FRANKFURT

U.K.	Pergamon Press Ltd., Headington Hill Hall, Oxford OX3 0BW, England
U.S.A.	Pergamon Press Inc., Maxwell House, Fairview Park, Elmsford, New York 10523, U.S.A.
CANADA	Pergamon of Canada, Suite 104, 150 Consumers Road, Willowdale, Ontario M2J 1P9, Canada
AUSTRALIA	Pergamon Press (Aust.) Pty. Ltd., P.O. Box 544, Potts Point, N.S.W. 2011, Australia
FRANCE	Pergamon Press SARL, 24 rue des Ecoles, 75240 Paris, Cedex 05, France
FEDERAL REPUBLIC OF GERMANY	Pergamon Press GmbH, 6242 Kronberg-Taunus, Pferdstrasse 1, Federal Republic of Germany

First published in Norwegian 1975
This Edition (Revised) 1979

British Library Cataloguing in Publication Data

Rock bolting.
1. Rock bolts
I. Schach, R II. Garshol, K III. Heltzen, A M
624'.15 TA722 79-41250

ISBN 0 08 022503 9

In order to make this volume available as economically and as rapidly as possible the authors' typescripts have been reproduced in their original forms. This method has its typographical limitations but it is hoped that they in no way distract the reader.

Printed in Great Britain by A. Wheaton & Co. Ltd., Exeter

CONTENTS

PREFACE

The Norwegian Institute for Rock Blasting Technique (IFF) have had stability and support of rock excavations as one of their research subjects. Rock bolting has formed an important part of this work.

Some of the results of the work have been presented at the yearly conferences on Rock Blasting (Fjellsprengningskonferanser), in articles in the technical press or included in technical books on the subject.

It has become clear that there is a need for a practical handbook on bolting.

Authorship and Editorship was mainly carried out by Schach and Garshol, whilst all the drawings were done by Schach.

A book such as this must be brought up to date at regular intervals. We would therefore be interested to receive comments on the contents, and details of practical experience of known or new bolt types.

Norwegian Institute for Rock
Blasting Techniques (IFF)

A. M. Heltzen
Director

INTRODUCTION

This book attempts to give an introduction to the practice and theory of rock bolting in mining and tunnelling. It is important that the reader is aware of the considerable differences which exist between mining and tunnelling. In the first case it is the excavated mass which is of interest, the voids formed being of secondary importance. The object is, however, to obtain an optimisation of the voids in order to give maximum ore recovery. In tunnelling it is the opening which is the primary object, while the excavated mass is not even of secondary importance except for the few occasions when it can be used for embankments or road-stone.

In mining, the progress must follow the ore. This often gives rise to stability problems. In tunnelling, there are much greater possibilities for selecting the line and cross-section of the tunnel in order to fit the topographical and geological features, while tunnels and cavities can be given a favourable siting and orientation which reduce the problems which can arise due to shearing zones, etc.

It is especially in the mines that rock bolting has had its greatest application, and first, and foremost in coal mines which normally lie in layered formations.

The first bolts were applied in a slate quarry in N.Wales in 1872. Bolting was introduced to the United States at about the turn of the century, although the new method of rock support was not received with any great interest. This was in the most part due to the conservative nature of the mining companies combined with a poor understanding of how the bolts really worked.

A further 50 years passed before bolting became accepted as a method of support. In Norway, A/S Sulitjelma Mines was one of the pioneers. With the so-called "sewing of the hanging wall" they introduced bolting as an economic and practical support method in the early post war years. From 1950 to 1955 bolting revolutionised a great deal of the American mining industry. By 1951, bolts had to a great extent replaced props in approximately 500 mines, 450 of them coal mines.

After bolting was accepted overall and introduced, there was a considerable reduction in accidents and simultaneously a large increase in production in the coal mining state of West Virginia. By eliminating props, etc., much improved ventilation was also achieved.

There is no doubt that many mines would have been forced to close down on both safety and economic grounds if bolting had not been introduced.

The pioneers have been the United States and South Africa. In the S. African mines, approximately eight and a half million bolts are used yearly, while the total for the USA is about 120 million.

Today, bolting is accepted as a very effective and relatively cheap support method, and props, which earlier were in general use, have almost disappeared. In mines, especially coal mines, there are few alternatives to bolting, but in tunnelling both shotcreting and insitu concrete can be applied. In general, steel and timber props are seldom used in Norway.

Even though bolting is of greatest importance in mining, this method of support is now applied for support of halls, tunnels, road and rail cuttings, dam foundations, etc. A number of these support systems are of a permanent nature, and the bolts are then usually grouted to prevent corrosion, also ensuring that any pre-tensioning of the bolts is maintained. In other cases immediate support is referred to and these bolts are often supplemented by other support methods at a later stage.

Bolting as a support method is fundamentally different from the conventional methods of timber props, steel arches, and various types of in-situ concreting. The conventional methods usually offer a passive support in that they first take up load when the rock begins to deform.

Therefore, fractures and cracks are allowed to open a small amount and the support system must take up part of the weight of the loose rock.

The factor which identifies bolts as a means of support, is that they are mounted in a borehole and therefore become elements within the rock, which are subjected to and react to the forces and deformations which occur. Bolts which are not tensioned work in a passive manner, even though they differ from other passive support systems.

When bolts are tensioned the support is active and is effective from the time the bolts are mounted. Deformation of the rock is reduced or prevented and fracturing is restrained. The normal forces on fracture planes are increased and the natural friction can be used to greater effect. Regardless of the bolt type there is an interaction between the bolt and the rock. This gives special advantages in coal mines where the settlement of the pillars is large yet where the hanging wall must be supported. Relatively small forces are required to retain the hanging wall in place, but the squashing of the pillars often causes overloading and destruction of props.

1. BOLT ELEMENTS

In this and the following four chapters an attempt is made to systemize and describe the existing bolt types and their characteristics, and also to deal with the individual components of the bolts. In principle, there are not so many types of bolts to choose between, but for a complete bolt there are many components, in different types and dimensions, and there are many manufacturers.

The possible combinations are therefore large, but the number is not so great in practice due to technical and economic reasons.

The possibilities which exist at present and which are technically and economically viable are not completely employed by the users.

BOLT STEEL

The round steel usually adapted for rock bolts used to be standard quality and dimension 19 mm (3/4") diameter, with threads cut at each end, used together with various types of expansion shells.

World wide a change has found place, toward better quality and smaller diameter.

As a replacement the most common diameters are round 16 mm and 18 mm in quality 60 to 70 kp/mm^2, and a yield stress of 40-50 kg/mm^2. With a higher breaking load, lower weight and acceptable prices, together with the fact that all threads are rolled, one can understand the points for the changing.

In addition to hot galvanized bolts, of which has been installed quite a number the last few years, bolts in stainless steel has been developed and presented on the Scandinavian market. The stainless steel, called Cromoround, is delivered in quality approximately DIN 1.4021, min. 13,5 % Cr., shiny. Breaking stress 65-80 kp/mm^2 and relatively extension 15-18%. Delivered as standard in dim. Ø 16 mm.

Also, bolts in crome/nickel steel, resistant against salt water and acid, are available. Same breaking load and extension as Cromoround.

For deformed rebar the main dimension is 20 mm diameter, but diameters 24, 30, etc. and up to 36 are available on the market, with breaking loads up to 50 Mp (tons).

Threading of deformed rebar increases costs as the combs must be turned off before a threading machine can cut the threads. These bolts are mostly installed in mortar (full bedded) or in polyester (anchorage in one end), for temporary support in black, and permanent support in hot galvanized or stainless steel. Also, deformed rebars are now available in stainless steel, called Cromorib, in the same quality as the round stainless steel.

The steel used and the production of Rock-Bolts is subject to a severe quality control. All deliveries should be guaranteed by a quality-certificate from the steel manufactures.

Yield and breaking load for rebar bolts and round steel

Bar Dia mm.		Rebar KS 40S metr.tons	Rebar KS 50S metr.tons	Rippentor 60 metr.tons	Round steel ST 70M metr.tons
16	Yield Break				14 18
18	Yield Break				18 23
20	Yield Break	12 18	15 22	19 24	22 28
24	Yield Break			29 34	
25	Yield Break	18 28	23 35		
32	Yield Break	30 45			

Table 1 shows the common bolt capacities.

BEARING PLATES

A multitude of bearing plates has been available threw the past years, square flat (the first on the rock), round, triangles of various types, rectangular, round spherical, plate with pressed sphere and slotted hole, and the latest known developed, the conical plate (called Fuji). There is no point in describing all the possibilities and varieties which have existed and are existing, see fig. 1.

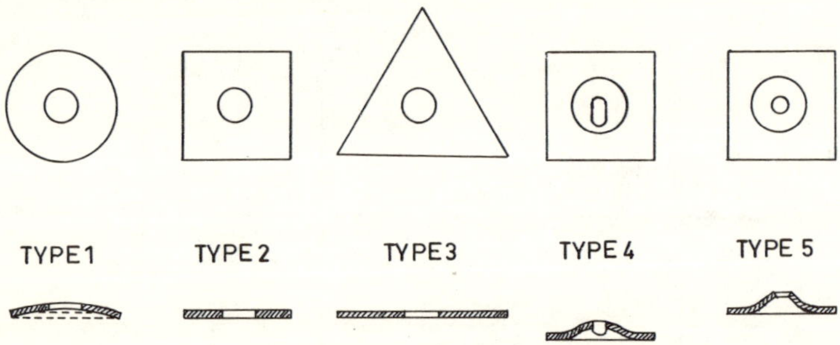

Fig. 1. Examples of bearing plates

Rockbolts for temporary support and rock-bolts combined for temporary and permanent support are always installed with bearing plates.

Rebar-bolts for permanent support, supplied in stainless steel or hot galvanized steel, will also provide bearing-plates if anchored in polyester or fully embedded in mortar in combination with accelerator.

In modern practice all such bolts are tensioned or prestressed to a greater or lesser extent, and there is no doubt that under most conditions the type of bearing plate chosen can be very significant.

The principle of pretensioning or prestressing a bolt is to stretch it between the anchor in the rock and the bearing plate at the rock surface. By systematic bolting, to prevent loose rock from moving or sliding from stable rock, it is necessary to use bearing plates, and the characteristic of the plate is supposed to be close to the characteristic of the bolt. The importance of the carrying capacity and the deformation of the plate has been very little estimated up to now.

What is the point in having an anchor taking a 30-40 tons load when the bearing plate is breaking at 10-15 tons?

Plates with bent up corners have shown that they can give a warning of failure, e.g. with spalling. Plane plates can give point contacts with the rock which can result in crushing and loss of load. Plates which are too thin will be pulled into the rock alongside the bolt and the corners have a tendency to bend out from the rock surface, fig. 3. Both tests and practice have shown that a very satisfactory type of plate is the round saucer shaped plate.

During mounting, the concave side is placed against the rock face and during tensioning of the bolt the plate is flattened against the rock at about 6,5 tons tension, fig.2,3,4. The result is a good contact with low surface pressure.

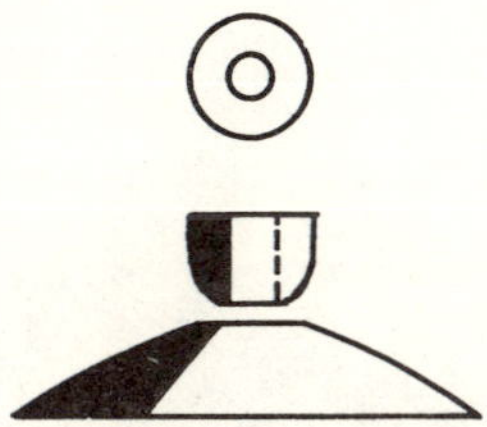

Fig. 2. spherical seating, type 1.

Fig.5 shows these types of bearing plates where the bolt is tensioned to 10 tons, giving also the calculated surface pressure in each case. The triangular plate should be avoided in soft rocks where the corners will crush the parts under them and bore

Fig. 3. Weak bearing plate with corners bearing out

into the rock. In harder rocks they should, however, be better. When blasting takes place nearby they will spring somewhat due to their shape, but the pressure on the contact surfaces is so large that extra crushing takes place and the final result is a reduction in the bolt tension dependent on the hardness of the rock.

See page 46 about tensioning of bolts. In order to ensure that the tension is maintained in a section bolted with expansion bolts and plane plates the minimum distance to a blast should be about 20 meters.

Fig. 4. Correctly tensioned bolt with round saucer shaped plate

The corresponding distance can be reduced to approx 10 m if round concave plates are used.

If polyester anchors are used the distances for the two types of plates can be reduced to 8 m and 3 m respectively.

These values are derived from tests carried out by the Norwegian Institute of Rock Blasting Techniques, but the distances mentioned are dependent on both the hardness of the rock and the size of the round. Naturally, bolting can be used right up to the face and often has to be, but the bolts should then be retensioned.

SURFACE PRESSURE UNDER THE SAME LOAD 10 TONS

Bearing plate type	1	2	3	4	5
Dimension mm	125 Ø	125 x 125	350 x 350 x 350	125 x 150	125 x 150
Thickness of plate mm	6	7	7	7	7
Contact area cm^2	28,48	23,57	23,57	40,25	44,76
Pressure kp/cm^2	351	424	424	248	223
Recommended load tons	10	8	8	14	18
	Crackings round the inner edge and the halfball had been pressed into the plate	All 4 corners were bent free from the rock	All 4 corners were bent free from the rock	All 4 corners were bent free from the rock	All 4 corners had contact with the rock and no marks on the nut or on the cone seat

Fig. 5. Difference in Bearing Plates Capacities

When strong spalling occurs in tunnels, large triangular plates with up to 60 cm sides are used, fig. 6. This ensures that a very considerable part of the roof is covered by the plates even with a normal bolting pattern. This has shown itself to be very effective and is one of the few conditions where the saucer shaped plate is not suitable.

Fig. 6. Large plates used aginst spalling

A new type of bearing plates, fig. 7, of Austrian

Fig. 7. GD-anchor plates

origin are used in connection with the new Austrian tunneling method (NATM). The square plate is pressed out in the middle to a sphere and given a slotted hole. It is necessary to use a washer under the nut.

The latest type of bearing plate on the market (May 1978) is the "Fuji" plate, see fig. 8.

It is a square or rectangular steelplate (ST.37 M) with a pressed cone and a conical hole in center.

The contact area on the nut is shaped as a half/ball, in order to get a complete contact between the nut and the conical opening in the plate all the way round, also for angular bolts.

Fig. 8. The GD-cone plate (Fuji)

At the same time the threads will be plastically pressed (forged), so that the bolt will be held not only by the threads, but also by friction.

With the conical plate one will achieve a higher maximum carrying load compared with other bearing plates with the same thickness of the plate.

The increase of the height of the nut also will contribute to a higher loading capacity, see fig. 8.

Rock Straps - wire mesh-roof ties.

Wire mesh and straps or bands are sometimes used in combination with bolts, usually as a working support right up to the face. The most usually applied wire mesh is woven type with 5 cm spacings and 2,7 mm thick wire, see fig. 9.

Fig. 9. Roof-ties and wire mesh

Rock straps are available in lengths of 2 meters, fig. 10, 10 cm wide and 5 mm thick; having a series of slots 7,5 x 2,5 cm at 15 cm centres along its length. Wire mesh is mainly used for catching small outfall and spalled flakes, but roof-ties can be used for the socalled "sewing", or suspending of large blocks, see fig. 11.

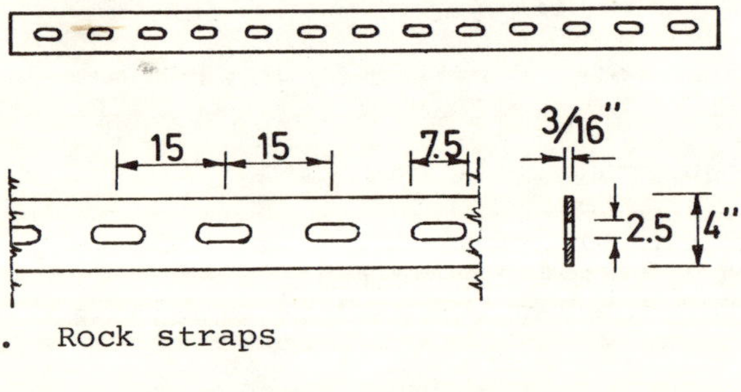

Fig. 10. Rock straps

Fig. 11. GD-rcof-ties

2. BOLT GROUP I - IMMEDIATE SUPPORT (WORKING SUPPORT)

For the grouping of bolt types the method used here is the same as that divised by technical principles.

A bolt for immediate support must be operative as soon as it is installed and must be practical and easy to work with. The use of a bolt as a working support often means that it will be mounted in the immediate neighbourhood of the face where it will be subject to additional forces due to further blasting.

All bolts in this group have a bearing plate and a rapidly applied anchor at the inner end and are normally tensioned. The various types differ from each other in the form of anchor used. Due to the danger of corrosion such bolts will only be considered as temporary support.

I A - SPLIT AND WEDGE

One of the first bolt types was the split and wedge, fig. 12. The bolt was split longitudinally for a few cms from one end and a steel wedge was inserted in the split. By hammering the bolt against the bottom of the borehold, the wedge was driven into the bolt which was therefore expanded against the wall of the hold. They are now little used in practice as they

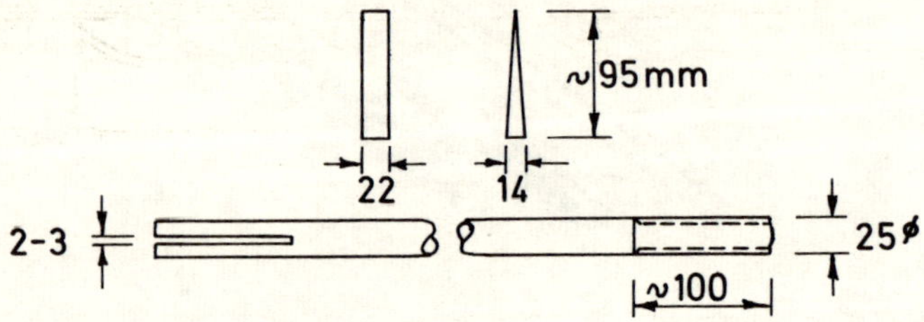

Fig. 12. split and wedge bolt

are somewhat uncertain in operation.

To a certain extent the method is used for eye bolts, anchor bolts, hangers etc.

The Kiruna bolt, fig. 13, is a special split and wedge bolt which still finds a good deal of use. This bolt has a cross type split and a suitable double wedge. This gives expansion in 2 planes and is therefore better than the single wedge type. Even so, its performance is not as good as the other types as far as safety is concerned.

I B - POLYESTER RESIN ANCHORS

At present there are two types on the market (in Norway), a German and an English. The German type comes in a glass tube, while the English type is supplied in a plastic cartridge, fig. 14.

The glass tube is filled with polyester and crushed quartz particles, the hardener is contained in a thin closed glass tube inside the outer tube. On inserting the bolt the glass is broken and in vertically upward holes a rubber collar should be used in order to prevent operators getting glass in the face. The hardening time of the German glass cartridges is from 10 minutes and upwards, and the storage life is several years.

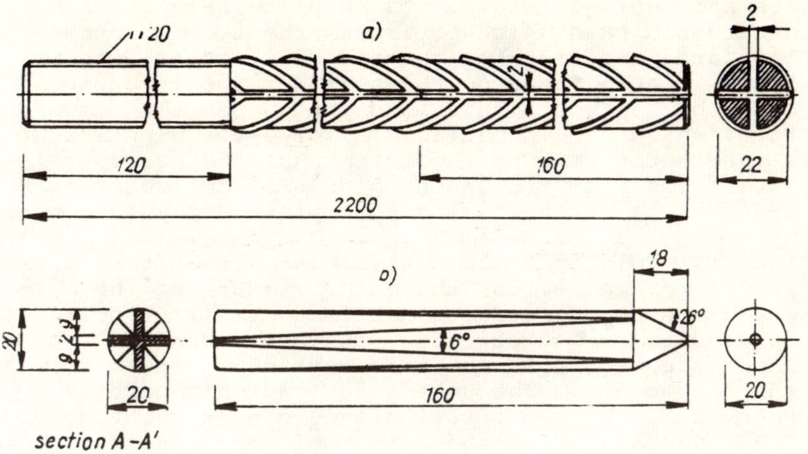

Fig. 13. Kiruna bolt

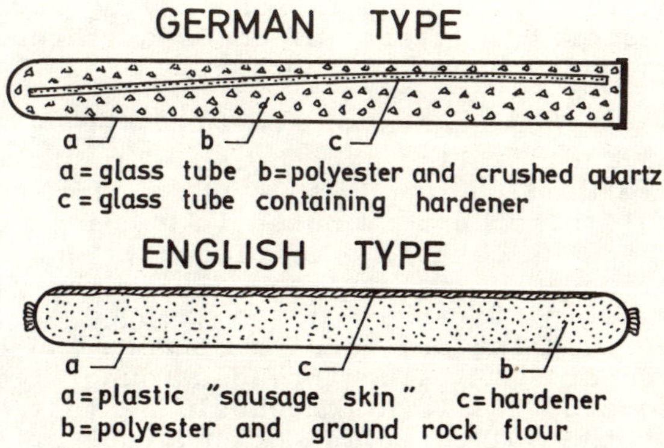

Fig 14. Polyester resin cartridges

The plastic cartridges are filled with a mixture of rock powder and polyester and the hardener lies between the polyester mass and the plastic skin.

As there is an interface between the polyester and

its hardener, a reaction takes place here forming a hardened skin which separates the two components. The hardening time for these English cartridges is from 2 minutes upwards and storage time is around 6 months, although this is dependent on the temperature, which should be low. A polyester type which has recently come on the market has the hardener in a thin plastic skin in the middle of the outer jacket and this has a shelf life of approximately 12 months.

The principle of these cartridges is that they are pushed to the back of the hole by means of the bolt which can be of round or deformed steel. A drill is coupled to the outer end of the bolt, which is fed in at the same time as it is rotated in order to mix the components. The speed of rotation should be about 120 rpm, and rotation should continue for about 20-30 seconds after the bottom of the hole is reached. After the required hardening time the bearing plate and nut can be mounted and the bolt tensioned. The hardening process also operates in wet conditions such that this method can be used under water. Tests show that the bolts require some keying surface in the polyester in the form of combs or threads in order to ensure a good anchorage.

As the cartridge has to be pressed against the hole bottom during rotation, the borehold depth must be closely related to the bolt length. The end of the bolt should also be terminated in a chisel edge to ensure that the cartridge is properly torn up. Tests with flat ended bolts have shown that the plastic skin was not perforated but was wrapped around the bolt so preventing any bonding between the bolt and the polyester. If using threaded round steel either the threads or the rotation should be left handed such that the polyester is not spun out of the hole. The adapter between the drill and the bolt is, in its simplest form, a nut welded to a 10 mm diameter rod. Bolts anchored by resins have been shown by tests to be better than expansion shells, particularly in softer rocks, fig. 15. It should, however, be noted that a longer anchor length is necessary in soft rocks than in the harder rocks.

In order to obtain optimum results from polyester

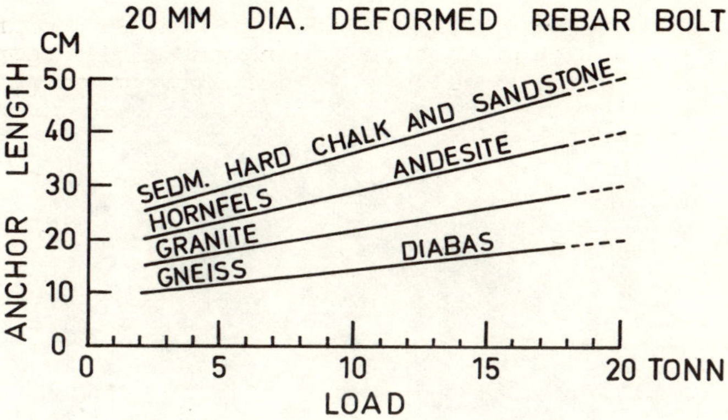

Fig. 15. Polyester resin anchor lengths

cartridges, it has been shown that the hole diameter should be 8-12 mm larger than the bolt diameter. It is also of great importance that the borehole is flushed of any drilling dust before installation of the bolt.

Field tests have shown that if the specified hardening times are observed, both the English and the German polyesters will enable loadings of 13.5 ton to be obtained with rebar bolts before yield of the bolts takes place. Failure occurs at the threaded section at loads of 18.5 to 20 ton, fig. 16.

Where conditions allow anchoring with a single polyester cartridge the cost would be approximately the same as an expansion shell.

GD-bolts, full mortar-embedded, prestressed with the GD-TOPAC-METHOD.

The GD-TOPAC-cartridge, a glasstube locked on both ends, contains an instantaneously acting accelerator, which has been developed specially for anchor-mortar and ceramic balls. The cartridge is placed at the end of the borehole and is embedded in the mortar.

Fig. 16. Tensile failure at threads

After complete filling of the borehole with mortar, the bolt is pushed into it, the cartridge is thereby destroyed.

The accelerator is mixed with the surrounding mortar and after a few seconds the mortar starts to harden at the end of the bolt.

The GD-anchor can be prestressed within a few hours, fig. 17.

I C - EXPANSION SHELLS

For working support with rock bolts, expansion shells are the most popular type both in Norway and overseas. Such a mechanical device is simple and reliable. These devices are found in many forms and from many suppliers but all fall into one of two categories; shell type or bail type, fig. 18. The figure shows only the principles of the two types as both of them

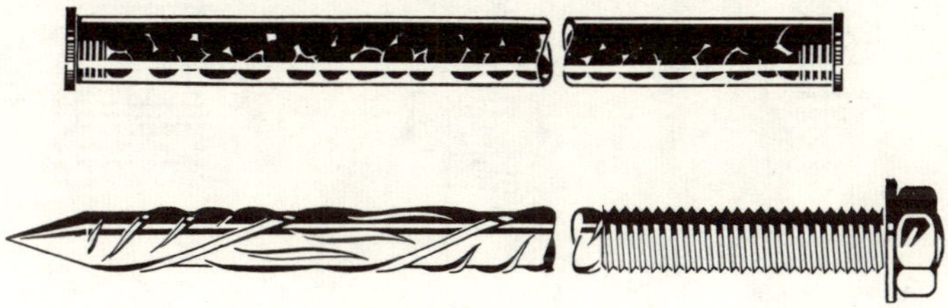

Fig. 17. GD-Topac-anchor

are available in many sizes for both different bolt and hole sizes.

They can also be made of cast iron or steel and of different lengths or other dimensions.

The principle for both types is that a conical nut is pressed between the toothed blades such that these are expanded against the hole wall.

So far, the method is the same as the split and wedge type, but the important difference is that the conical nut (wedge) is captive on the threads of the bolt such that on loading, the nut is pulled harder between the blades. Thus, provided the shell is not

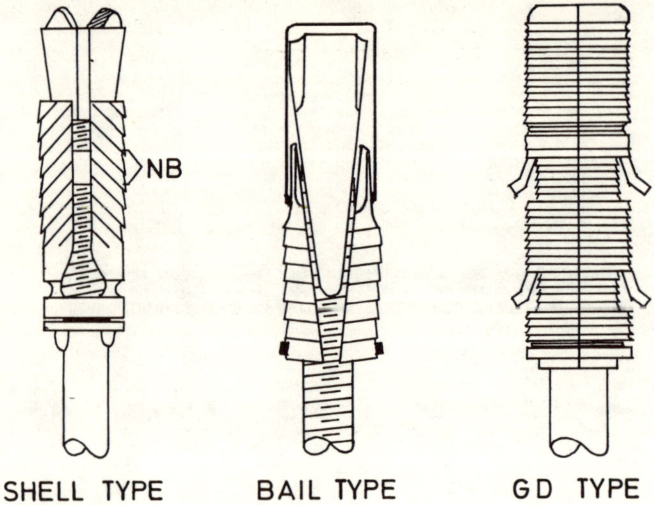

Fig. 18. Expansion shells

overloaded, the fastening is improved with increased tension on the bolt. As seen in fig. 18, the shell type has its collar at the lower end while the blades are open at the top.

The blades stand on a collar or a pair of forged ears on the bolt which prevent the blades being pulled down the bolt. The nut has two edges which fit in the gap between the blades such that the shell and the nut rotate together. By screwing the nut and shell together by hand the nut is drawn down the threads. When the shell stops on the collar, the nut is drawn in between the blades. A suitable expansion to the hole size can thus be achieved and the bolt inserted in the hole. The further expansion and fastening can then be achieved in 3 ways:

1. The ears on the nut are pressed against the hole bottom. When the bolt is rotated the nut remains stationary and the shell is pressed upwards against the nut and expands against the hole wall.

2. The bolt does not reach to the hole bottom. The shell must then be opened sufficiently by hand before insertion such that the blades have contact and friction against the rock such that the bolt can be rotated while the nut and shell remain stationary.

3. The shell can have a thin elastic steel spring placed between the blades, see NB on fig. 18. The ends of the spring catch in the hole wall and hold the shell and nut in place as in point 2.

When the shell has come into contact with the rock by one of the above methods the further expansion will be achieved by the tension in the bolt. If shell type expansion bolts are being used the steel spring will be found necessary if the insertion is not to be problematic. The shell itself is also closely dependent on the borehold diameter. This dependence means that the holes can be tight when using the longer drills of a series or worn drills. On the other hand, the bolts are difficult to insert and uncertain in operation if the hole diameter is too large.

The bail type anchor, as shown in fig. 18, is composed of two loose blades which are fastened to a steel bail which goes over the end of the conical nut. For insertion, the blades can be bent out slightly and they will then slide along the hole wall without hindering the insertion of the bolt, fig. 19. One can then begin to rotate the bolt so the nut is held by the bail which is retained in place by the blades laying along the borehold wall. The blades are then further expanded. It is also possible to directly pull the bolt without rotation after it is placed in the hole as the nut will be pulled into the blades.

Special types of expansion shells are also found, e.g. tandem type i.e. two shells in series intended for especially loose rock, fig. 20. Of the standard types there are, of course, some which are better suited to particular conditions but it is difficult to determine this beforehand. The best

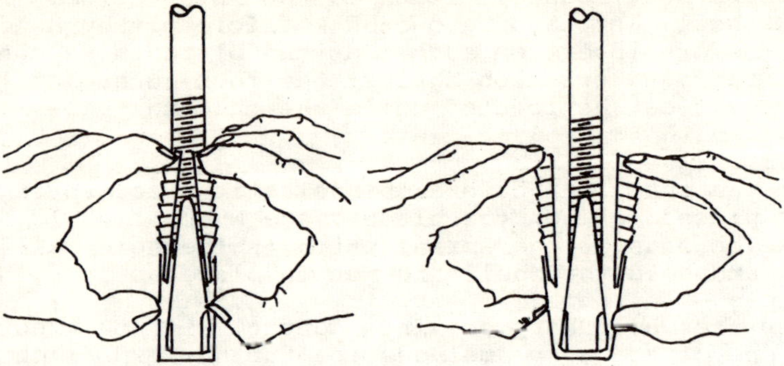

Fig. 19. Expansion of a bail type before mounting

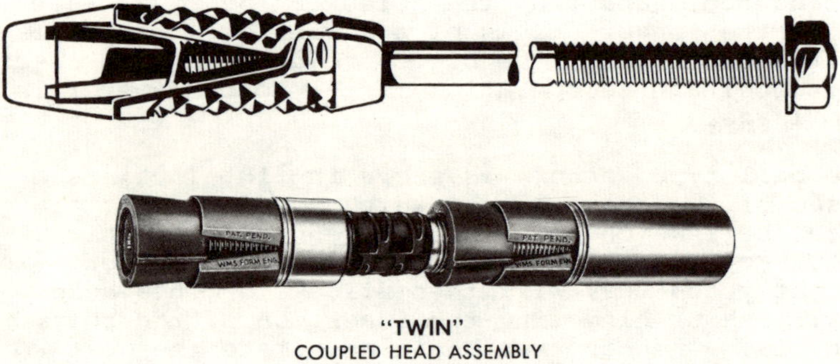

Fig. 20. Expansion shells for big holes

method of finding the most suitable type is to undertake direct pull-out tests on site. Normally all the standard types are fully acceptable and it is non-technical grounds which often determine the choice of type.

Special types of expansion shells of malleable iron, with high capacity for poor rock and wider boreholes, are made with + 20 mm expansion and the shell is locked at maximal expansion and cannot slide over the wedge (cone), fig. 20.

A special type of expansion shell (GD-rockbolt) is made of a plastic compound, fig. 21. The wedge is forged with a collar and a coneshaped head on the bolt. The advantage of this type is its simple installation without rotating the bolt and its great load capacity. The slightest pull expands the shells and this makes it possible to use with short bolts, endhooks for hanging up ventilation pipes, etc., fig. 22.

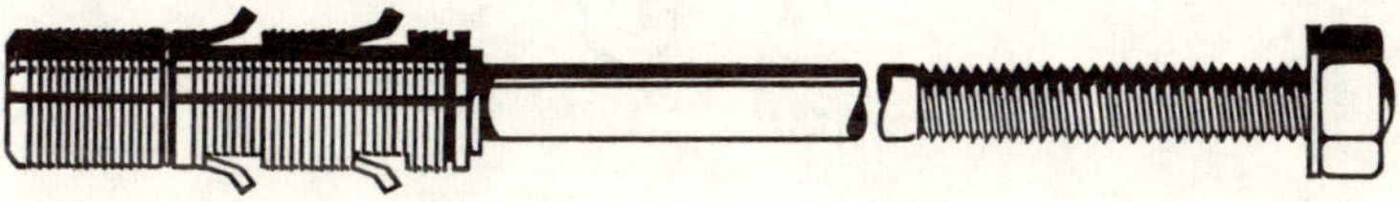

Fig. 21. GD-Bolt

3. BOLT GROUP II - PERMANENT SUPPORT

Whether a bolt can be considered permanent or not is dependent on what length of time one thinks of and how quickly the bolt would be expected to be destroyed by corrosion. In this field there are unfortunately few details available, and while there are examples of unprotected bolts being undamaged after several years, there are also examples of the reverse. It is clear that the conditions for corrosion vary a great deal from place to place. Practice has developed to mean that a permanent bolt is a bolt grouted with cement mortar.

It is recommended to use dry ready-mixed mortar, delivered in watertight bags, just to add the foresaid amount of water. The different types vary mainly in the method of filling the mortar. The bolts used are normally deformed rebar without any other anchorage than the mortar and either with or without plates. Grouted bolts are not normally tensioned even if they have bearing plates. Such bolts cannot be used where immediate support is required as the cement grout takes so long to cure.

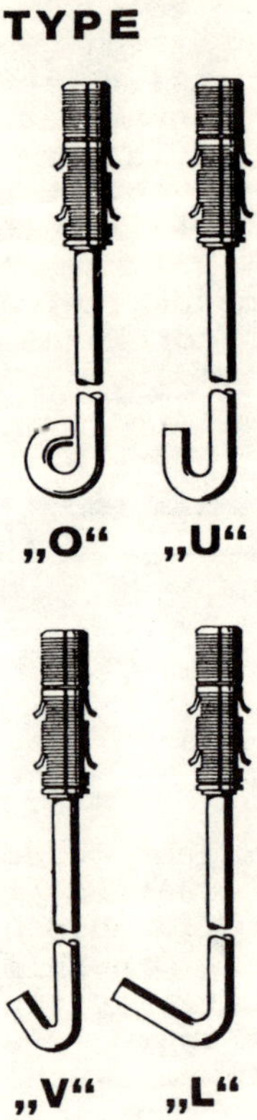

Fig. 22. Endhooks

Also, cement grout is likely to be damaged by nearby blasting particularly when it is curing. In tender documents in the U.S.A., it is often included that grouting of bolts with cement grout cannot take place before the face is more than 40-50 meters distant.

By using quick setting cement or additives of accelerators the hardening time can be considerably reduced, but the bond between the grout and the bolt will nevertheless be damaged by too small a distance from a blast.

Grouted bolts which are subjected to detonation loading can be overloaded as the movement of the free rock face is not divided over the whole length of the bolt but is concentrated at fractures or at the surface itself. Even if the bolt itself is not damaged the grout will be damaged at these points.

Bolts with plates and nuts which are anchored only with cement grout should not be tensioned. This is because the load is transferred only to a very limited length of the bolt near the rock surface, which results in the grout loosening or cracking. It also develops tensile forces in the rock near the surface which can easily lead to tensile cracks forming. Any possible tensioning must be divided over the whole length of the bolt by anchoring it at the inner end and tensioning before the grout begins to harden.

The normal grout used is sand/cement in ratios between 50/50 and 60/40. It is necessary to take great care with the mixing of the grout. The sand must be sized such that it is well graded with a maximum grain size of 2 mm. After the cement and sand have been well mixed and the water is then added, the mass must be carefully worked. Herein lies the secret of a good working result. In order to obtain a plastic grout, bentonite, up to 2% of the cement weight, can be added. Alternatively, prepacked commercial grouts which have an expansion property can be used (Betokem EXM Embeco or equivalent). In vertically upward holes the water quantity should be carefully controlled such that the grout is pumpable. If the mixture is too wet it will run out of the hole.

II A THE PERFO METHOD

This method uses a thin perforated metal tube, split longitudinally and with a diameter a little smaller than the borehole. The two halves are filled separately with mortar and they are then joined together by binding at the ends with soft iron wire, fig. 23. The filled perforated tube is then inserted in the borehole. When the rebar bolt is pushed out through the holes, it fills the annulus between the tube and the borehole wall with grout. This type of bolt can be equipped with plate and nut for the support of the rock surface. If the rebar is not threaded it is sometimes possible to use a plate held by a wire clip.

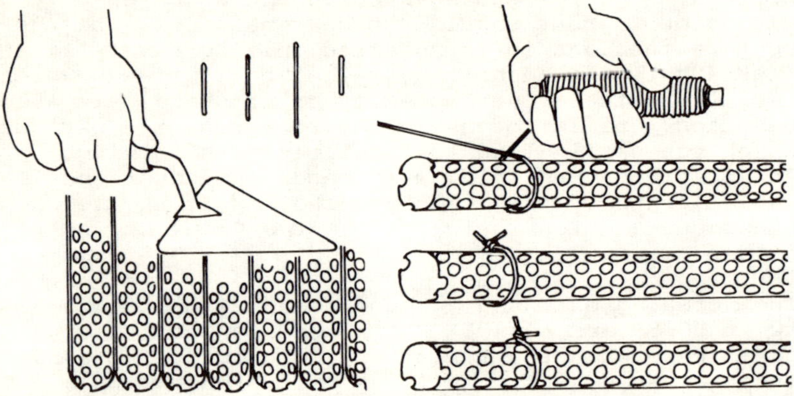

Fig. 23. Filling of perfo tubes

II B BERGJET METHOD

The Bergjet bolt has derived its name from the Bergjet apparatus for pumping cement grout in a borehole, fig. 24,25,26. The apparatus consists of a container made up of an upturned cone with a hemisphere on the top. In the hemisphere there is an internal cover which seals against the opening with a rubber packing piece with the help of compressed air. The compressed air is added at the top of the container with a pressure of approx. 2-4 kp/cm^2. The air pressure can be regulated by a reduction valve and the pressure is read off from a manometer. At the bottom of the container there is an albow with a connector to a hose which is fed to the bottom of the borehole. The container is filled with grout which is then forced out

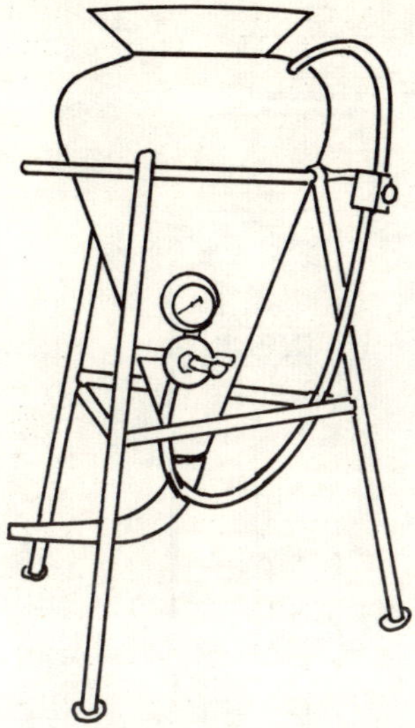

Fig. 24. The Bergjet pump

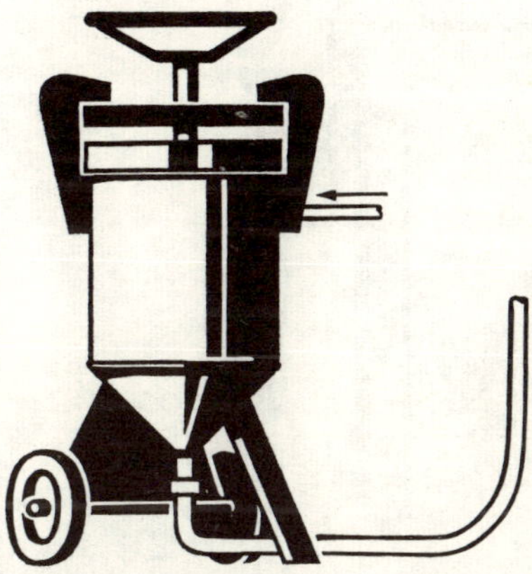

Fig. 25. GD-Mortar pump

through the hose by air pressure. The hose is pushed in to the bottom of the borehole and is pulled out in small jerks, such that it always is in contact with the filled grout. When the hole is so full of

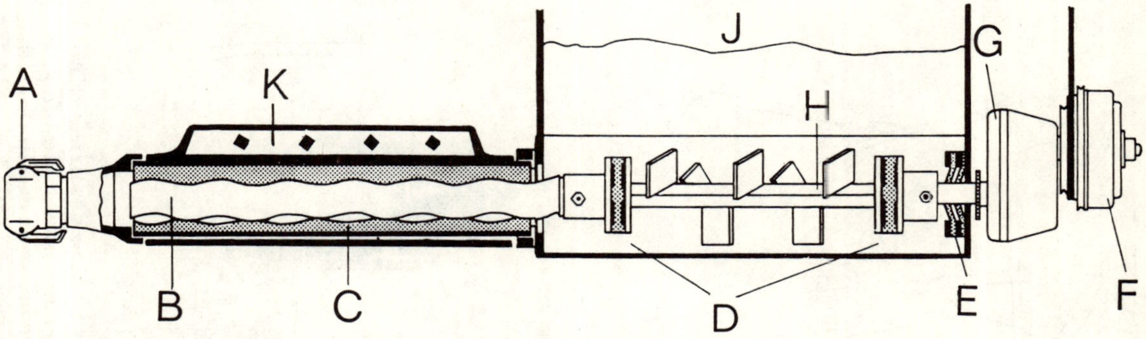

The PUTZMEISTER P 11 worm type pump with adjustable stator, diagram view.

A	PUTZMEISTER quick-action coupling	F	Pulley with clutch
B	Worm	G	Gearbox
C	Adjustable stator	H	Agitator
D	Flexible drive shaft couplings	J	Mortar hopper
E	Shaft seal	K	Adjustable clamp

Fig. 26. Putzmeister mortar pump

grout that the grout and the bolt together will fill the hole the air is turned off, the hose withdrawn and the bolt pushed home. The bolt is now installed and becomes operative as the grout hardens. In vertical holes it is advantageous to fasten the bolt with a wedge to prevent its creeping down and also to pack around it with cotton waste, rock wool, etc., fig. 28.

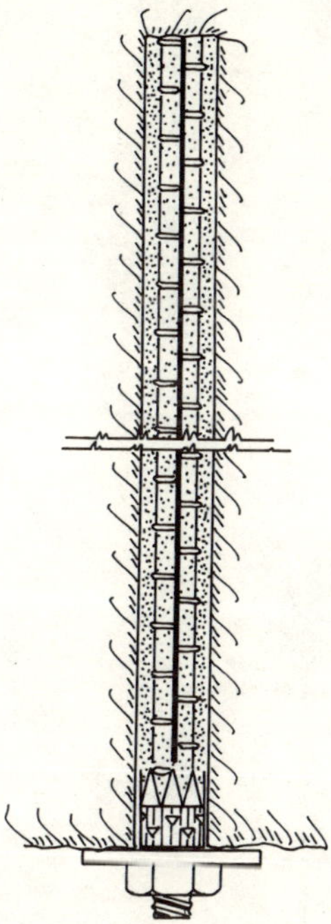

Fig. 27. Injecto bolt

II C INJEKTO METHOD

In the Bergjet method the hose from the compressed air device is passed to the back of the borehole. In the Injekto method the hose is only pushed 10 - 15 cm into the hole through a thin sheet metal fitting, fig. 27. A thin semi-stiff plastic tube is inserted to the hole bottom for air bleeding. When one is certain that a sufficient amount of grout has been pumped in, the hose is removed and the bolt inserted. In this method a softer consistency of grout is used than for the Bergjet method.

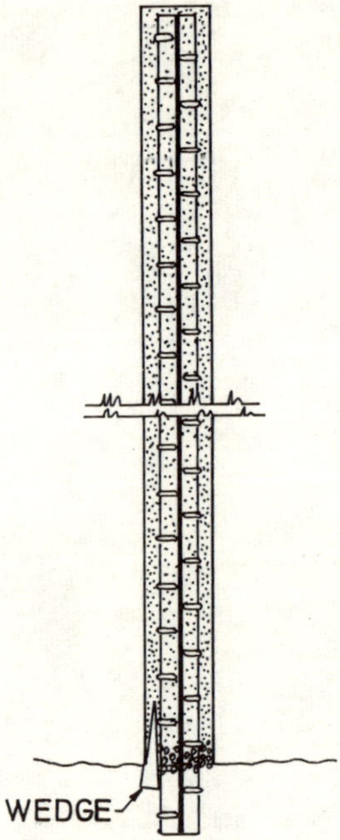

Fig. 28. The Bergjet bolt

With the GD-TOPAC method, where an accelerator in a glass tube at the end of the borehole makes the mortar quick hardening, while in the remaining length the mortar is still soft, it is possible to prestress the bolt shortly after installation, so that we get a prestressed, fully mortarembedded bolt, fig. 29. This type is specially suitable for forepoling, see page 85.

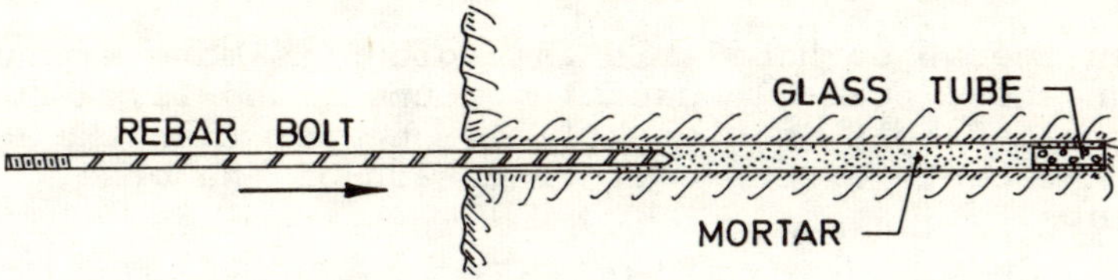

Fig. 29. GD- Topac method

II D - POLYESTER GROUTING

The use of polyester (see II B) for anchoring and rust protection is technically a very good solution, but grouting of the whole bolt length with polyester is at present too expensive even though special large, slow-hardening cartridges, which are somewhat cheaper, are available. There can also be practical problems in rotating the bolt through a long length of polyester cartridges. The manufacturers are at present looking at the possibility of reducing the viscosity of the slower hardening cartridges as it is to be expected, that the installation technique becomes easier.

4. BOLT GROUP III - COMBINED IMMEDIATE AND PERMANENT SUPPORT

Bolts in this group must have immediate and reliable operation and be equipped with a bearing plate. Normally they should be capable of being tensioned. In order to satisfy the requirements of permanent support there are numerous possibilities and it is the client's responsibility to select a method on the basis of the cited requirements.

It is apparent that there is relatively little knowledge of this group. Many examples have been recorded where the working support has been carried out with unprotected bolts, and permanent bolts have then been mounted completely separate. This is expensive for the client and can also be delaying and problematic for the contractor.

Even though each bolt in this group is more expensive than those normally used in groups I and II, a ready installed bolt in group III is cheaper than the sum of the costs of one bolt from each of the other groups.

Bolts in this group can obviously be also applied away from the face as a pure permanent support.

Bolts in this group are lately supplied in stainless steel or hot galvenized steel in both round - and rebar steel instead of black steel. When using wire mesh and shotcrete black bolts also can be recommended.

III A - EXPANSION BOLTS IN STAINLESS/HOT GALVANIZED STEEL WITH NONE CORROSION SHELLS.

These types of bolts are used for immediate support in constructions where bolt support is wanted to be permanent. These bolts give an immediate support by tensioning. The bolts are installed with a plate and a nut, see fig. 18.

III B - REBAR AND ROUND BOLTS WITH PADDLES IN STAINLESS STEEL OR HOT GALVANIZED INSTALLED IN RESIN OR MORTAR.

With these types of pin point anchored bolts in polyester resin there is no need to protect the bolt completely when using stainless steel. If the bolts are pin point anchored in mortar, we recommend an accelerator in the bottom of the hole, see fig. 29, 30.

III C - HOLLOW EXPANSION BOLTS.

This type of bolt is used for immediate support in constructions in bad rock where bolt support is to be permanent. The bolt combines both these functions. It gives an immediate support and can be tensioned.

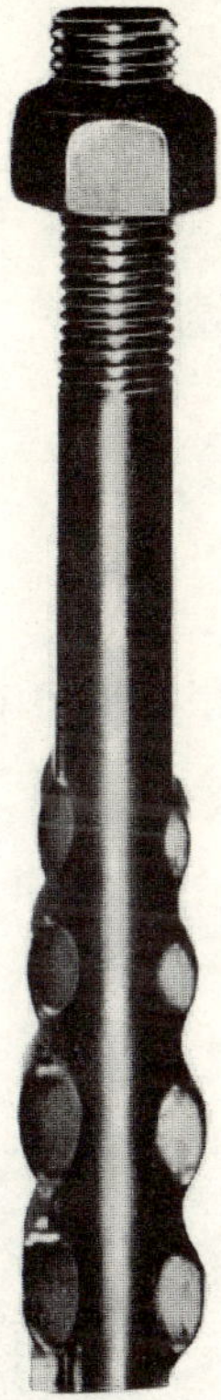

Fig. 30. Round paddle bolt

The bolt is installed with a plate and nut. The bolt can then be made permanent by grouting and post-tensioned when the face has advanced a convenient distance see fig. 31, 32.

III D - Normal Deformed Rebar Bolt with Resin Anchorage.

Bolts as described in I B, polyester resin anchors, can also satisfy the requirements for permanent support is, as outlined in II D, a rapid hardening cartridge is used at the hole bottom and slower cartridges are used for the rest of the hole. The bolt is thus

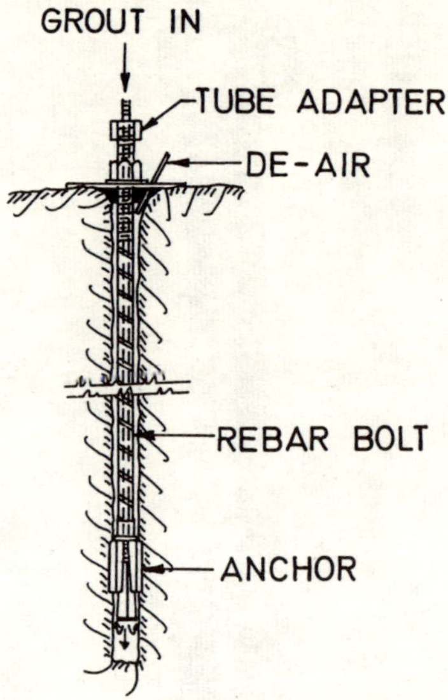

Fig. 31. Hollow expansion bolt

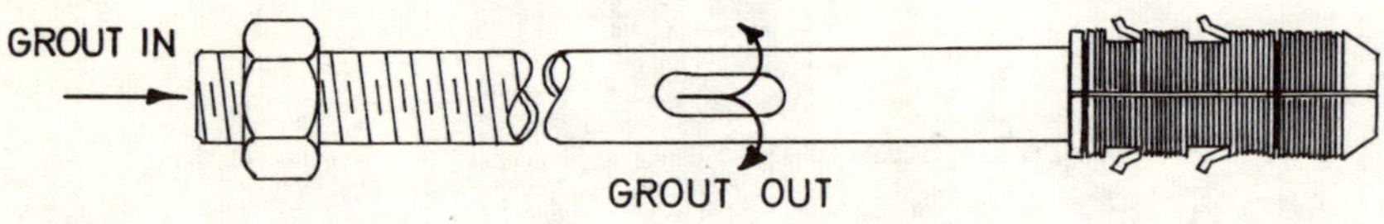

Fig. 32. Combi bolt for 45 mm hole, hollow for grouting

completely covered by polyester when mounted and is therefore protected. The hardening time for the slower cartridges is about 2 hours. After the inner cartridge(s) is hardened, the bolt is tensioned to the desired load and a pre-tensioned, corrosion protected, permanent bolt is achieved.

Previously combined expansions bolts were grouted by using grouting pipes and air outlets, see fig. 33, and split and wedge bolts installed in mortar (Kiruna

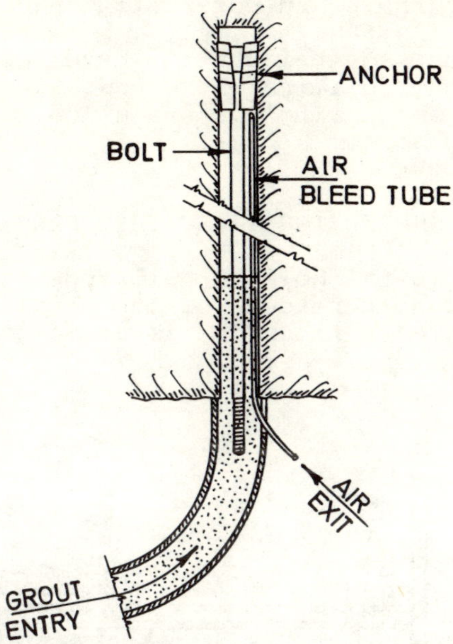

Fig. 33. Grouting of a normal expansion bolt

type, fig. 13).

These types are now too expensive to install in competition with the more timesaving bolts mentioned above.

5. BOLT GROUP IV - SHUTTERING BOLTS

In most large underground excavations where technical installations require relatively dry rooms, contact concreting to the rock wall is used. All contractors have their own method of fastening the shuttering but the most usual is a deformed rebar bolt cement grouted into a hole drilled into the rock. An expansive shuttering bar is then welded to this and both are lost. Also, there is a minimum delay of 2 days after mounting the bolt before the cement grout can be loaded. This method is commonly used for walls in tunnels and power station halls.

In recent times, several systems have been introduced which either eliminate or reduce the loss of the expensive shuttering bar and allow immediate loading of the bolt.

IV A - SHUTTERING TIE WITH EXPANSION SHELL

For fastening in the hole, a bail type expansion shell with a cone threaded for the particular tie bar steel is used. Insertion is as follows, see fig. 34, 35.

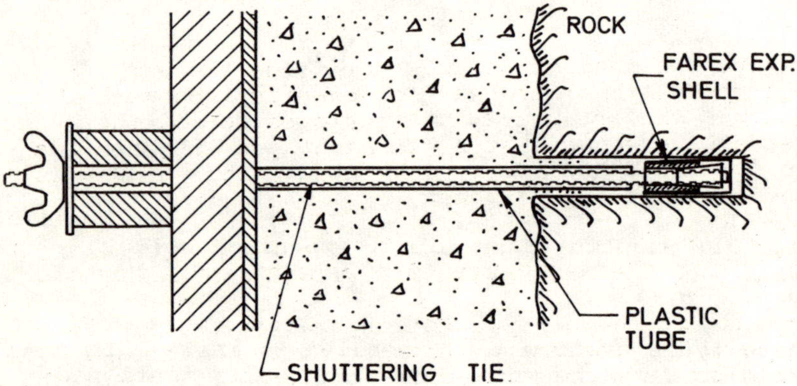

Fig. 34. Shuttering tie with expansion shell

The bar is screwed into the expansion shell, which is then inserted into the hole. The bar is then pulled by hand until it fastens.

A tension of about 3 tons must be applied so that the cone is pulled well into the shell and causes the latter to grip the hole wall. If this is not done the shuttering will give 1-2 cm under the concrete pressure. Tensioning is best achieved with a suitably shaped lever which operates on the wing nut.

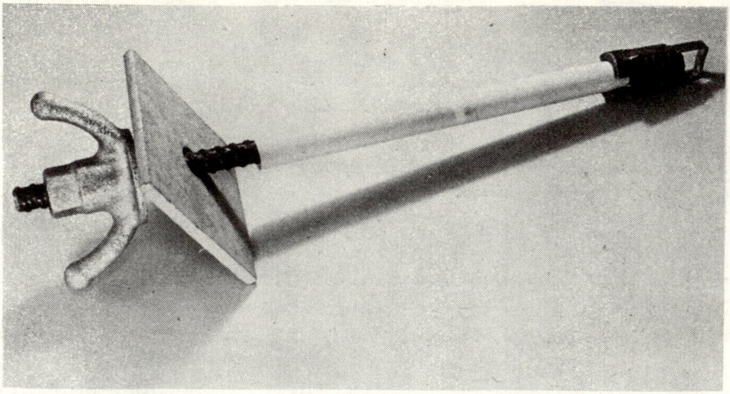

Fig. 35. Model of a shuttering tie with expansion shell

Before the shuttering is erected the tie bar is equipped with a cardboard or plastic tube such that the bar can afterwards be retrieved for further use. Tests have shown that a tie bar which has been loaded to 10 tons can be removed by hand. This method is the simplest and cheapest.

IV B - SHUTTERING TIE WITH RESIN ANCHOR

In this method a rapid setting polyester cartridge is used for fastening the bolt in the borehole. The mounting is as described in Section I B. A special nut, covered by a polystyrene foam tube, is screwed halflength onto the bolt. From the other side, through the shuttering, a short tie screws into the other side of the special nut, see fig. 36. The length of the bolt is important in that the outer face of the special nut should be at the inner face of the shuttering board. Tensioning of the tie is by the same method as type IV A. When the shuttering is stripped off the outer tie is recovered but that inside the nut is lost. The special nut can also be recovered by the use of a special key, fig. 36.

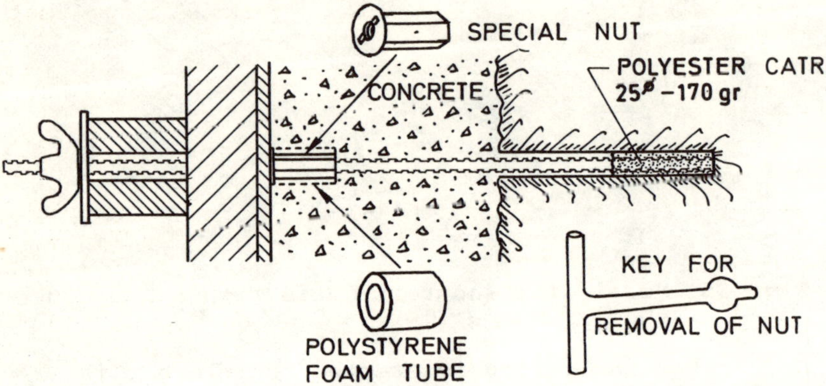

Fig. 36. Shuttering tie combined with deformed rebar fastened with cement or polyester resin

6. BOLT GROUP V - YIELDING BOLTS

It has often been a desire in securing work in mines and tunnels to have a bolt which has the ability to slide outwards in the borehole and at the same time keep the tension in the bolt.

Ordinary rock bolts yield and extend up to 18 % and then they break.

Considerable larger extension would be desireable in certain cases especially in road building where it is necessary, for instance, to keep the consolidation under control.

The two following types of bolts solve these problems:

V A - SPLIT SET (DEVELOPED BY INGERSOLL-RAND COMPANY AND DR. JAMES J. SCOTT; INVENTOR).

Split-Set is a friction roof bolt consisting of a long steel tube with a slot through its entire length, fig. 37.

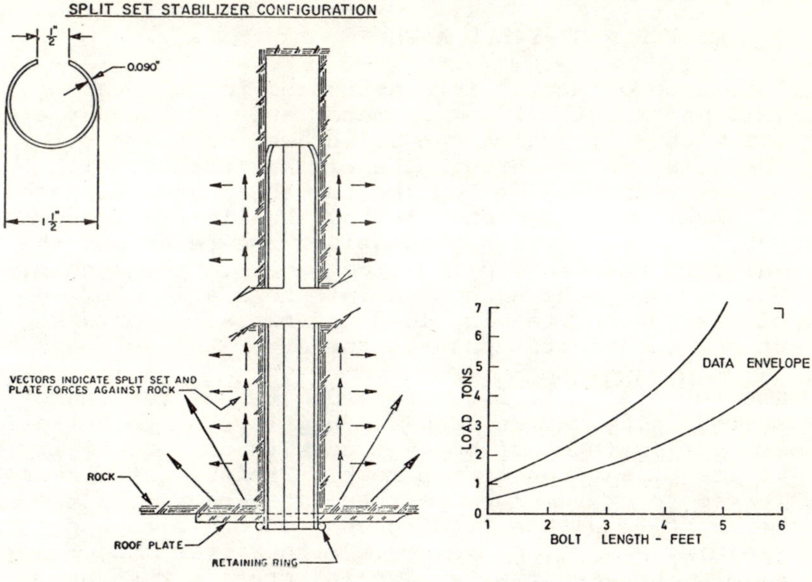

Fig. 37. Split Set construction and slit load

The split tube is forced into a hole which is smaller in diameter than the tube. In fact, the tube is plastically deformed, with the borehole serving as a die to form the final shape right in the roof. The split tube maintains intimate contact with the wall of the hole and imposes a distributed radial loading. This radial loading provides the source of frictional resistance to movement of the rock. Additional fractions of Split Set include a flange on the lower end and an expansion anchor at the top.

The friction roof bolts have an 1½ in. outside diameter and are normally installed in 1 3/8 in. holes. Depending on hole smoothness, diameter throughout the hole length, straightness, and frictional coefficient, an anchorage of 3/4 to 1,5 tons per. ft. is obtained initially.

V B - GD-YIELD ANCHOR

This type of anchor is constructed for a rebar (Rippentor) with 24 mm diameter and consists of a cup with a lid which can be screwed on the cup. There is a hole through the lid and the cup with diameter 29 mm. Inside the cup there are 64 pieces of steel balls. When screwing the lid tight on the cup, the steel balls jam against the rebar and the walls of the cup and lock the rebar to the anchor. When loaded, the anchor can withhold a load of 90% of the rebars breaking load and can slide outward on the rebar with constant force.

The two sketches, fig. 38, indicate how the anchor successfully can be used. A highway was to be built across a stretch of boggy land with no possibility to use piles down to the rock. Prefabricated reinforced concrete slabs were held together with the rebar bolts with a tight anchor on one side and a yielding anchor on the other with a free end of 1 meter. Between the two "walls" crushed rock and gravel were filled in and loaded trucks drove forwards and backwards for a certain time to stabilize and consolidate the ballast. After a period of 3 months the packing was completed and only infinitesimal settings were observed.

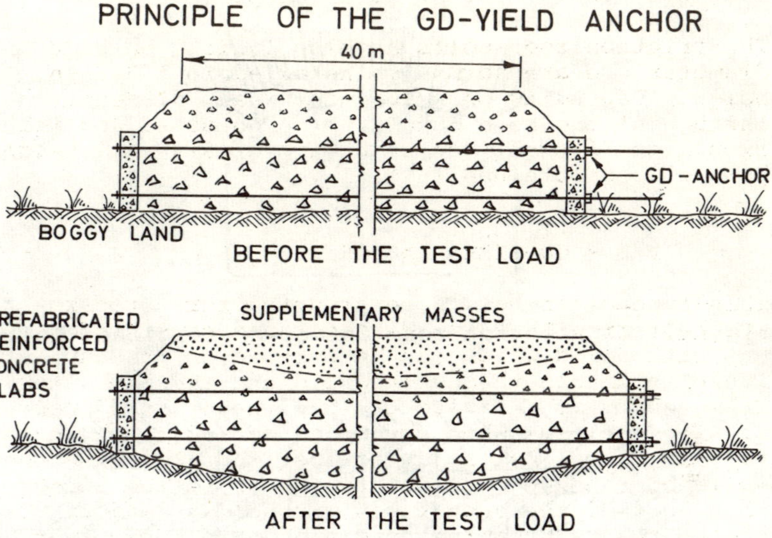

Fig. 38. GD-Yield anchor

7. PRETENSIONING OF BOLTS

The principle of pretensioning a bolt is that the bolt is stretched between the anchor in the rock and the plate at the rock surface. By using systematic bolting a preloaded zone is created over the cavity which will work as an arch and be able to take a considerable load from above, see fig. 39. By single bolts, which are correctly placed and installed, loose blocks can be tied back to stable rock and also weak stratified rock can be riveted together to form a beam. fig. 40. Fig 41 shows how to obtain the best effect of individual bolts. Under normal conditions the bolt should be tensioned to 50%

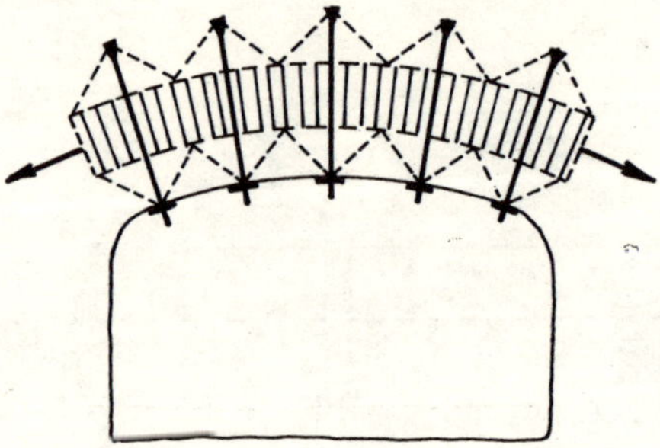

Fig. 39. Compression zone formed by tensioned bolts

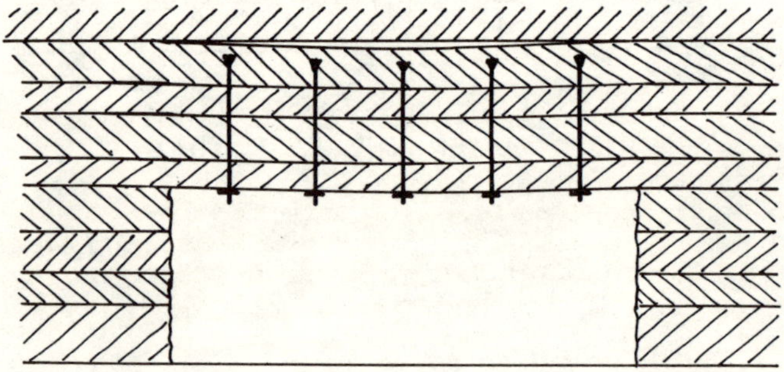

Fig. 40. Beam effect

of its yield strength for permanent support and 70% for working support. In practice, this gives values for standard quality 19 mm steel (37 kg/mm^2) of 4 tons for permanent and 5,2 tons for working bolts.

TENSIONING WITH ADJUSTABLE SPANNER

The most usual method of tensioning a bolt is to use a large adjustable spanner. With this method only a limited torque can be applied to the nut, depending on for example, the operator's position with respect

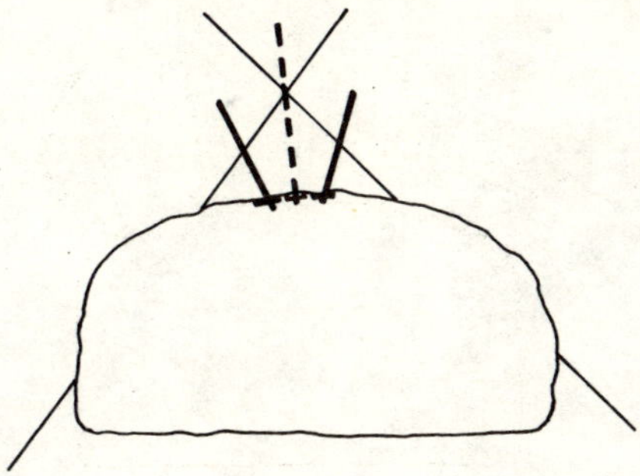

Fig. 41. Single bolts support

to the bolt. Therefore a piece of pipe is often used to increase the length of the lever and a torque of 15-20 kp m. equivalent to approximately 3-4 tons tension, can be achieved. The method is not accurate enough if it is stipulated that all bolts should have the same tension.

Tests on bolts tensioned by adjustable spanner have shown values between 1,6 and 3,5 tons; the difference being mainly due to the varying friction in the threads and between the nut and the spherical seating.

TENSIONING BY TORQUE WRENCH

A better method for tensioning is to use a torque wrench where the torque can be read off in kp m or lbft. By this means all the nuts can be given the same torque but due to the friction in the threads and against the seating the bolts will still not have the same tension. The variation is, however, less than when using an adjustable spanner.

TENSIONING BY COMPRESSED AIR WRENCH

Several types of compressed air wrenches for torqueing nuts are available, see fig. 42. On most of them it

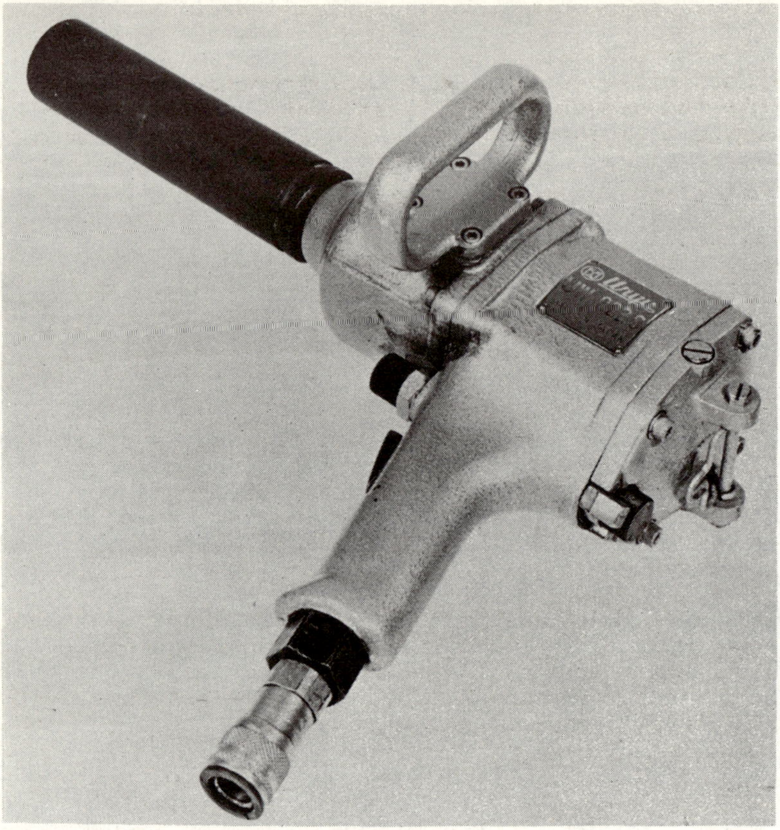

Fig. 42. Compressed air torque wrench

is possible to preset the desired torque and the machines can give a much higher level than the two previously described. The higher torque naturally gives a higher bolt tension but the variation in tension due to frictional effects is still not eliminated.

As examples, the following three c/a wrenches designed for rock bolting can be mentioned:

Ingersoll-Rand impact wrench IR-5340-To torque 20-76.

Gardner-Denver impact wrench W-6 torque 20-45 kpm.
Atlas Copco impact wrench LMS44-HR-01, 04 torque 10-45 kpm.

With constant use, such impact wrenches tend to slip from the preset value due to wear. The equipment should therefore be calibrated after about each 200 applications. Fig. 43 shows a graph relating tension to torque.

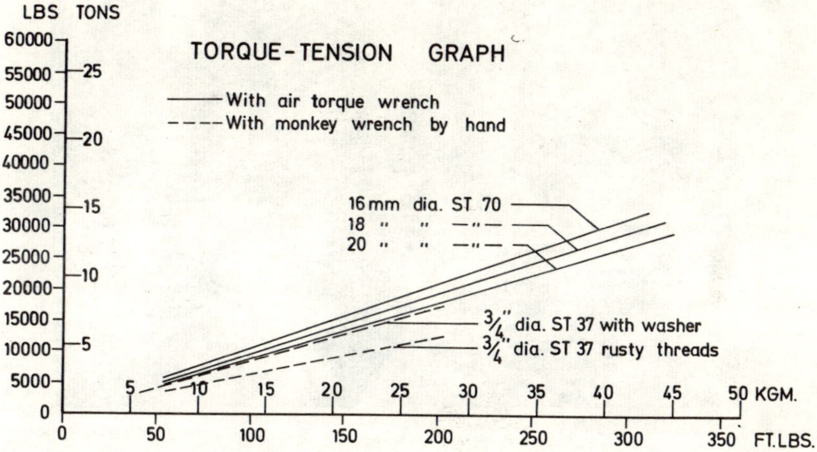

Fig. 43. Tensioning table

TENSIONING WITH HYDRAULIC TENSIONERS

The most reliable method of tensioning is, without doubt, the hydraulic tensioner, which applies a direct pull to the bolt without torque. This means that friction is eliminated. The tensioner is shown in fig. 44 and 45, and consists of a through axle with a hand wheel at one end and a female threaded chuck at the other. A small hydraulic jack surrounds the axle just above the handwheel. The jack is equipped with an indicator pin which jumps out when the preset required pull is reached. Insertion and tensioning of a bolt proceed as follows:

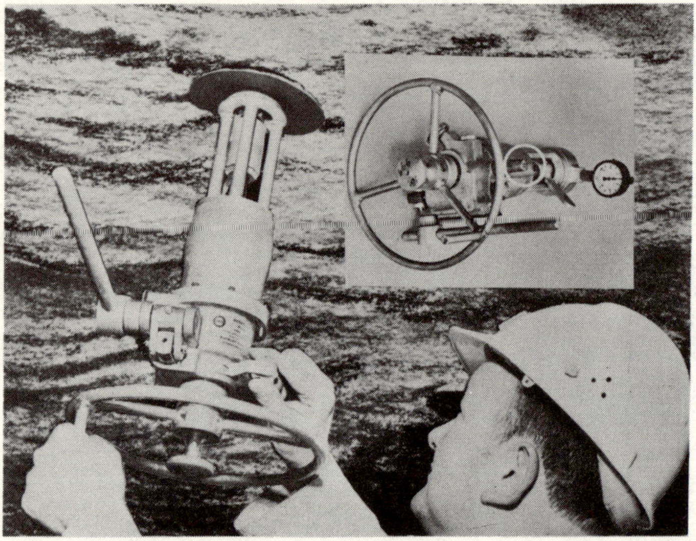

Fig. 44. Hydraulic bolt tensioning device

The expansion shell is attached to one end of the bolt and the plate and nut on the other such that about 1 cm of thread remains outside the nut. A spherical seating should be used in order to obtain an axial pull. The bolt is then attached to the tensioner and inserted into the hole. If a bail type expansion shell is used and it was expanded somewhat before insertion, the bolt will now hang in the hole. This is particularly important in vertical holes. The slack which exists is taken up by rotating the hand wheel until the expansion shell has a fast grip, the last turn should be a double-handed turn.

The valve on the pump is then closed and the jack is pumped up, the bolt being stretched without tension. When the indicator pin jumps out the nut is screwed

Fig. 45. Detail of hydraulic bolt tensioner

up finger tight against the plate or seating. The integral pipe spanner with handle is then used for tightening of the nut. The pump valve is opened and the tensioner is removed from the bolt by rotating the handwheel in the reverse direction.

The indicator pin is usually adjusted to 6 tons by the supplier. When the valve is opened the bolt tension will be reduced by about 500 kg, leaving the bolt tension at 5,5 tons. The pin can be adjusted up to 10 tons if required, fig. 44.

The tensioner is easy to use and gives almost equal tension to all bolts. It can also be used to determine the tension in previously installed bolts, as one can follow the loading in tons on a gauge and read off the maximum value when the nut loosens. The tensioner described and illustrated is the Elbroc rock stud tensioner mark 8 designed for 10 tons tension. A larger type, mark 10 can develop up to 30 tons tension.

8. DETERMINATION OF BOLT LENGTHS AND BOLT PATTERNS

LIMITED LOOSE BLOCKS

Functionally, this is the simplest support problem for the use of bolts. Normally, it will be possible to determine the size of the block by observing the position and directions of the fracture planes defining the block. For large blocks it is useful to have a compass with an inclinometer to register the block. and dip of these planes and then sketch the block. On the basis of these observations, the bolt length can be determined.

The bolt should reach 1-2 meters into solid rock. The length should be determined taking into consideration the degree of certainly with which the block size was determined. If possible the actual position of the observed fractures should be controlled when drilling the bolt hole.

$L \geqq D + 1,0$ (all dimensions in meters) see fig. 46.

The bolting pattern or, in this case, the number of bolts is dependent on the weight of the block. When the block dimensions are known the volume can be calculated and by multiplying by the specific gravity (normally 2,6 - 2,8) we obtain its weight.

Normally, shear strength of the fractures is assumed to be zero so that if it does exist we have an extra safety margin. When the bolt type has been determined the number of bolts can be calculated. It is often assumed that only half the bolts are effective so the number is doubled giving a safety factor of 2,0. For the particular application one should consider the safety factor demanded and the accuracy of the basic assumptions. Quality control of the bolting process is also important, e.g., when using expansion bolts the number of effective bolts will be greater using a bolt tensioner than it would be using an adjustable spanner.

A block such as described can be so large or alternatively so unstable that it is unsafe to drill holes directly into it. The bolts must then be placed alongside the block and metal straps fastened across the block. The assessment method will be the same as before but because of the altered directions of the bolts and the difficulty of evening out the load the safety factor should be increased.

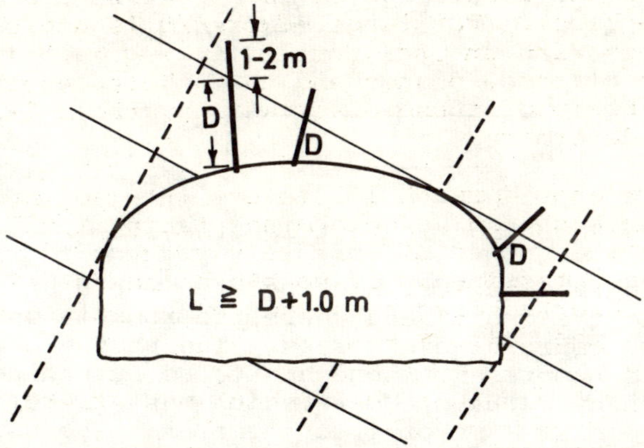

Fig. 46. Bolt lengths for limited loose blocks

SYSTEMATIC BOLTING NOT TENSIONED

In a tunnel such as is shown in fig. 47, where it has been found necessary to use systematic bolting, the length of the bolts in the center of the drive can be determined approximately from the following I.F.F. formula:

(1) $L = 1{,}40 + 0{,}184\ B$

Here, B is the tunnel width in meters. In the derivation of this equation a circular or semi-circular form of the roof has been assumed.

It was also assumed that the blasting out of the tunnel, would give a readjustment of the stresses such

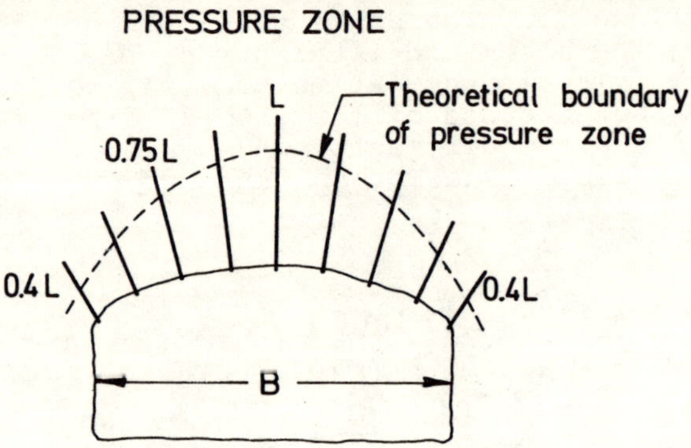

Fig. 47. Reduction in bolt lengths

that a parabolic self bearing arch would be formed spanning the tunnel and where the rock is only subject to compressive stresses. The unstable rock below this compression zone has to be suspended from the arch by means of bolts which must therefore extend at least 1 meter into the arch. The bolts can then be reduced in length towards the outer walls. In practice, shorter bolts will only be used when approaching the walls in very broad tunnels. The graphical table, fig. 48, shows bolt lengths for systematic bolting as advocated by several authors, while fig. 49 shows examples of systematic bolting with grouted bolts.

SYSTEMATIC BOLTING - PRETENSIONED

In many conditions it is desirable to be able to establish a stressed arch or beam over the tunnel. With the bolts normally used in Norway today, it is possible, with good tensioning, to give the rock a compressive stress of about 0,5 kp/cm^2. This presupposes the rock is almost stress free; if this is not so the stress from the bolts must be added to the existing value by superposition. Thus the possibility of the bolts making a significant alteration to the

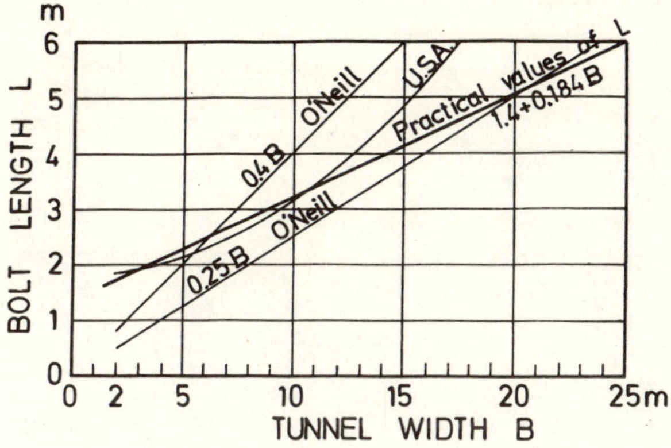

Fig. 48. Dimensioning curves for bolt lengths

stress picture is very limited. The use of bolts to create a stressed zone is therefore restricted to stress relieved rock or laminated or bedded rock.

For determining the recommended bolt lengths, fig. 47, or the following formula can be used.

(1) $L = 1{,}40 + 0{,}184\ B$ I.F.F.
(2) $L = 1{,}60 + \sqrt{1{,}0 + 0{,}012\ B^2}$ Jorstad

All the bolts should be of the same length, and as seen from the formula (1) and (2)-fig. 48, there is some difference in the bolt length dependent on the formula adopted.

This is due to the different test methods and theoretical backgrounds for the formula and on the different authors interpretations. The two formulas have been selected from the multitude available in order to show the approximate variation in bolt length within which we feel we should operate. The chapter on principles gives further information on how bolt lengths should be selected for individual applications.

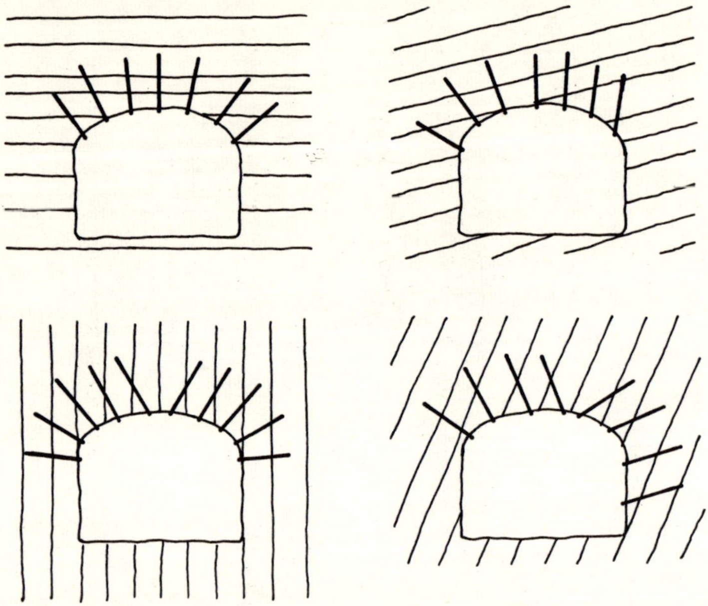

Fig. 49. Examples of systematic bolting

For such bolting applications the effect of the variation of the relationship between bolt length (L) and distance between bolts (a) is very important.

In order to establish a compression zone it is necessary that the ratio L/a should be almost 2,0. This should always be aimed for, but local conditions of fracture spacing etc. may have to be given special consideration and it may not be achieved, see fig. 50.

The criterion L/a - 2,0 has been found by tests carried out in several places. The tests are based on the tensioning of bolts inserted in photoelastic plastic sheets where the isochromatic fringes give a picture of the stresses in the plate in polarised light.

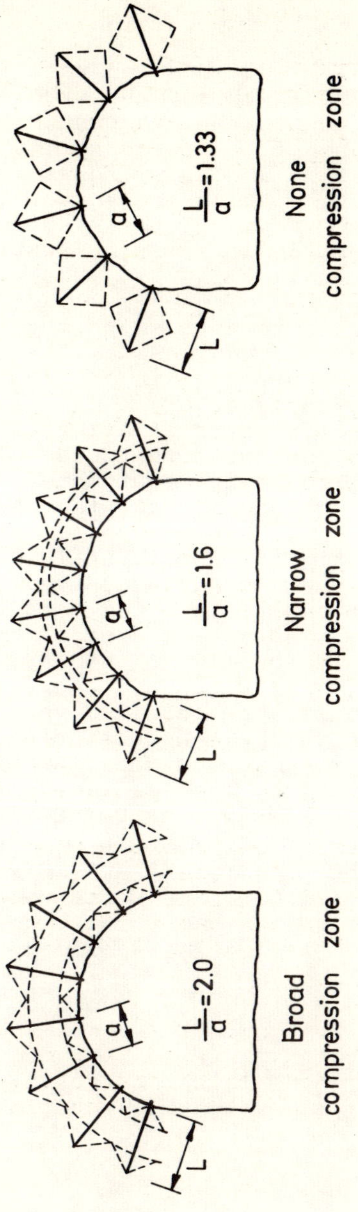

Fig. 50. Establishing a prestressed arch

SUPPORT OF WALLS AND CUTTINGS

In blasting work, high cuttings and walls in high cavities are often met with and the stabilisation of these can entail a considerable amount of work depending on the structure of the rock and its fracture frequency. It is often necessary to use long bolts of large diameter. The following paragraphs give a number of examples of the necessary calculations and considerations to be done for bolt stabilisation of a high rock cutting.

In all the examples it is assumed that a relatively steeply dipping plane of weakness exist with a strike parallel to the longitudinal axis of the cutting. Both smooth and rough planes are considered in the calculations and coefficients of friction of 0 and 0,5 are assumed for the two conditions. The bolts are assumed to be grouted 26 mm diameter deformed rebar and the working load is half the breaking load.

In the first case, fig. 51, horizontal bolts have been used while in the second case, fig. 52, the bolts were inserted at an angle. The wedge of rock outside the plane of weakness has its center of gravity at point T. As shown in the figure the forces are resolved into components parallel and at right angles to the weakness plane. The parallel vector P is further resolved in the bolt direction P_H and P_B. In the cases where friction on the plane is assumed the force P is reduced by the frictional force F to give R which is then resolved as before for P_F. Numerical values are given in brackets on the figures in order for specimen calculations of bolt numbers to be undertaken.

Common Assumptions (example in brackets)

Fall angle α (57^c)		
Bolt angle β (6^c = 1:11)		
Safety factor s = 2		
Friction coefficient f = 0.5		
Specific gravity of rock	γ t/m^3	(2,80)
$G = \frac{1}{2} \cdot H \cdot B \cdot \gamma$	t/m	(112,00)
$N = G \cos\alpha$	t/m	(70,00)
$P = G \sin\alpha$	t/m	(87,36)

$P_H = P/\cos\alpha$	t/m	(139,77)
$P_B = P/\cos(\alpha-\beta)$	t/m	(125,51)
$F = Nf$	t/m	(35,00)
$R = P - F$	t/m	(52,36)

$\sin 57^{c} = 0{,}780$
$\cos 57^{c} = 0{,}625$
$\cos 51^{c} = 0{,}696$

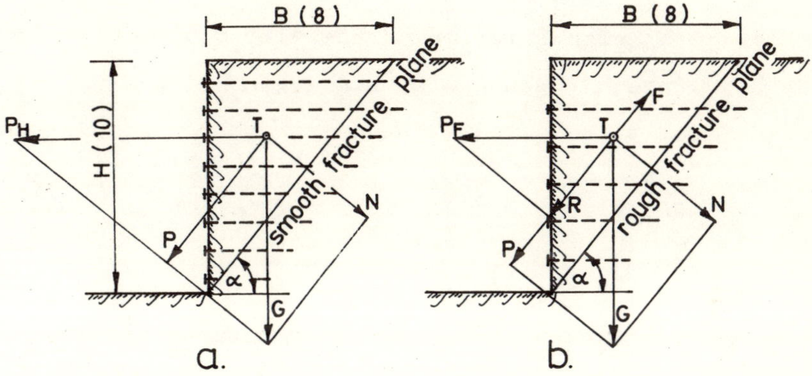

Fig. 51. Support of a cutting with grouted bolts

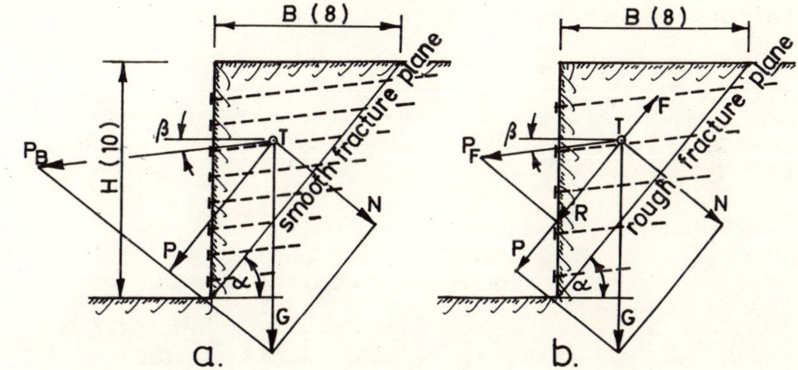

Fig. 52. Support of a cutting with grouted bolts

It can be assumed that the bolts will have as high a shearing strength as their tensile strength.

Fig. 51 a, no friction

$$P_H = \frac{P}{\cos \alpha} = G \cdot \frac{\sin \alpha}{\cos \alpha} = \frac{1}{2} \cdot \frac{H \cdot B \cdot \gamma \sin \alpha}{\cos \alpha}$$

Fig. 51 b, with friction on the slip plane

$$R = P - F = G \cdot \sin \alpha - Nf = G \cdot \sin \alpha - G \cdot \cos \alpha \cdot f$$

$$= \tfrac{1}{2} H \cdot B \cdot \gamma \sin \alpha - \tfrac{1}{2} H \cdot B \gamma \cos \alpha \cdot f$$

$$P_F = \frac{R}{\cos \alpha}$$

Fig. 52 a, no friction

$$P_B = \frac{P}{\cos (\alpha-\beta)} = \frac{G \sin \alpha}{\cos (\alpha-\beta)} = \frac{\tfrac{1}{2} H \cdot B \cdot \gamma \cdot \sin \alpha}{\cos (\alpha \ \beta)}$$

Fig. 52 b, with friction on the slip plane

$$R = P - F = G \sin \alpha - Nf = G \sin \alpha - G \cdot \cos \alpha \cdot f$$

$$= \tfrac{1}{2} \cdot H \cdot B \cdot \gamma \cdot \sin \alpha - \tfrac{1}{2} \cdot H \cdot B \cdot \gamma \cdot \cos \alpha \cdot f.$$

$$P_F = \frac{R}{\cos (\alpha-\beta)}$$

BOLTING PATTERNS

Bolting patterns cover the distance between bolts, the geometrical pattern of their placing and their direction. In the case of the clearly defined loose block the length of the bolts should be determined and the individual bolts placings should be marked on the block surface with its direction also indicated, fig. 54. In systematic bolting it is often

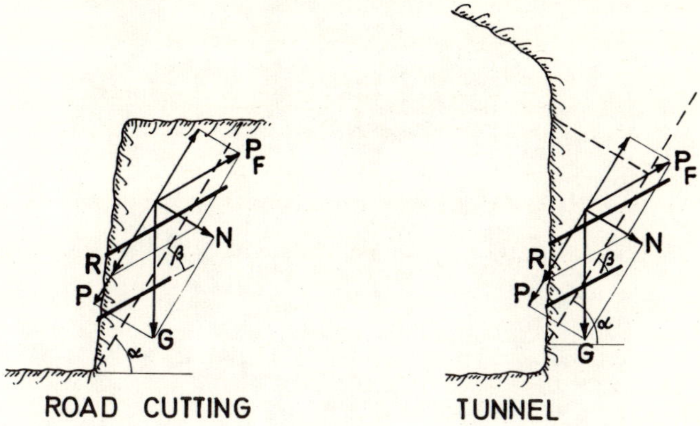

Fig. 53. Examples of support

desirable and simplifies the work of the driller, that each individual hole is marked in position and direction before the driller commences, see fig. 55.

In deriving a general pattern for systematic bolting the first thing that must be decided is the distance between the bolts. This is done on the basis of the average size of block in the roof (a function of the fracture frequency) and the usual shape of the blocks (quadratic, slender etc.). The distance is adjusted such that it is less, in all directions, than the average fracture distance for the main fracture systems.

As the bolt distance determines the number of bolts, the cost and the possible loads should also be considered. If the fracture frequency indicates that the bolt spacing should be less than 1 meter then the use of rock straps and/or wire mesh is required.

In practice, most underground excavations are much linger in one direction than the others such that one normally determines the distance between the rows (at right angles to the longitudinal axis) and the number of bolts or bolt distance in each row. Further,

Fig. 54. Example of bolt marking

the pattern should be chosen either rectangular or diagonal, see fig. 56. It should also be stated whether the bolts should be at right angles in the rock surface or whether other considerations necessitate a different direction. For a full description the exact description of the bolt type, the dimensions and the installation procedure should be included.

Example

Distance between rows	r	2.0 meters
Distance between bolts	b	1,2 meters
Pattern		diagonal
Bolt length		2,0 meters

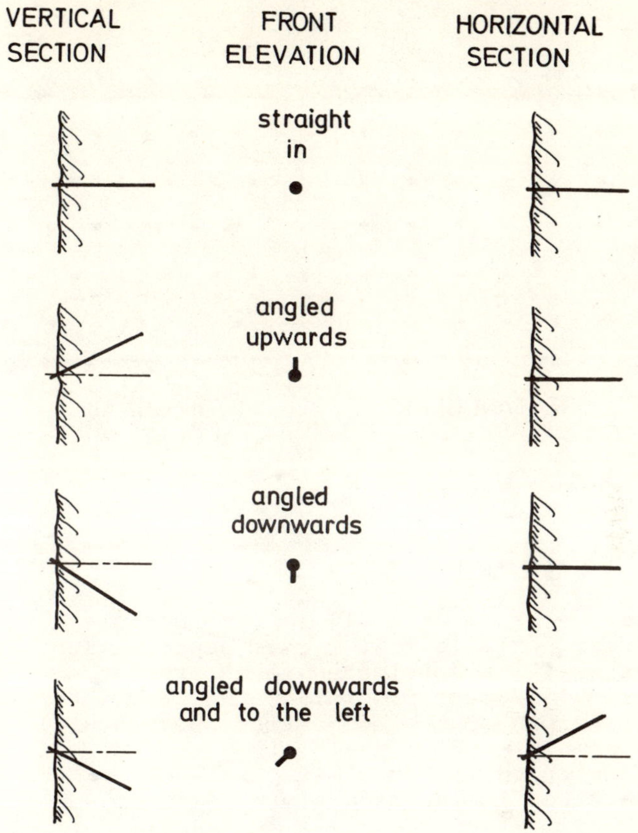

Fig. 55. Bolt marking

Bolt type	3/4 " expansion shell bail type spherical seating saucer shaped plate.
Mounting direction	At right angles to rock surface
Bolt tension	5.5 tons
Rythm	2 rounds/bolting cycle

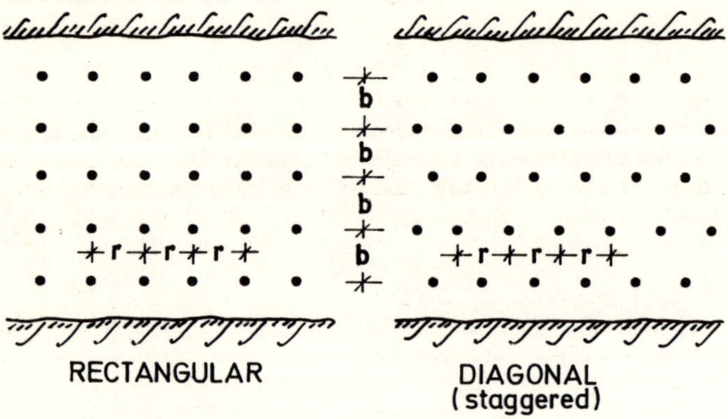

Fig. 56. Bolting patters

9. PRINCIPLES

In this chapter, we shall look more closely at the background principles for rock support with bolts, as far as it is considered necessary for solving individual support problems. The many theoretical discertations on the subject of bolting will, therefore, remain mostly outside the scope of this handbook and those with a greater interest in the subject are referred to the literature.

The way in which a bolt works as a rock support is subject to several theories, depending on which bolt type is under consideration. For systematic bolting of the roof with pretensioned bolts it is often considered that the bolts bind the rock layers together to form a self supporting rock beam. Particularly in fairly flat lying rock beds this happens as the bolts in principle should not take more load than their pretension, the latter increasing the friction between the elements so that they bear their own weight like a beam. It is more usual to consider bolting as a means of creating a compression zone in the rock such that this pressure arch works as a load bearing construction.

The difference between these two methods is in reality not so great. The use of single bolts in isolation areas is seen as an anchoring of a surface fractured rock section, or a block, to the solid and stable rock lying behind it.

If the bolts are not tensioned, but grouted, they will work as a reinforcement of the fracture rock and a retension of the friction and stability conditions which already exist such that possible time dependent factors are prevented from causing instability.

Tensioned expansion bolts set up concentrated stresses at the anchor and the plates. When these are used systematically, it has been shown, both by model studies and full scale tests, that in unfavourable conditions cracks and fractures spanning between anchors can be formed. This happens particularly when the roof is flat, all the bolts are of the same length and the rock has a flat lying foliation or plane of weakness. The compression zone which is obtained between the plate and the anchor is of the order of 0.5 kp/cm^2, even with a pretension of 5-6 tons on bolts set in a lm x lm pattern. As one can see, the stress in the well renowned pressure zone is relatively small. When choosing a bolting system, many differing formulae can be found in the literature for calculating bolt length and bolt distance. The formulae are in most cases based on the span of the excavation and the observed fracture spacing. Some of these formulae can, under some circumstances, give extreme values such that experience and common sense must be exercised in each individual case.

A further point, that should be remembered when considering the theoretical and practical aspects of a bolting scheme, is the safety factor of the rock against immediate spalling. As spalling has in many cases not occured and the rock is stable enough for holes to be drilled in it, then its factor of safety is 1.0 or greater. Primarily bolting is to increase the safety factor against immediate fall of rock. The consideration of time dependent effects will be discussed later.

The object of these introductory remarks on the theoretical background of support of rock by bolting was mainly to indicate that a universal model or theory for the method of operation of bolts cannot be given. The theoretical considerations work as a basis and reference for experimental work and also for practical bolt assessment, but most practical bolting problems require additional consideration and it is these latter which are the determining factors for the specific system for that site. We shall therefore discuss the problems of bolt support from the practical and rock mechanics angles on the basis of what can be observed in the tunnel, as it is here that the solution to the problem most often lies.

When one is faced with a stability problem in a rock cavity, one must first and foremost decide on the type of problem occurring. Is it rock spalling ? Is it clay zones ? Is it blocky rock, etc.? Here it is important that the time and the available equipment which is necessary must be used in order to obtain the best results as this is the basis for all further considerations. One must not hesitate to insist on having close access to a high roof or adequate lighting, etc., in order to make a satisfactory examination. Very often it is extremely useful to observe, at close hand, the scaling down. In some difficult cases it is necessary to hose down the rock surface in order to examine it properly. A hammer or pinch bar to tap with are useful tools for detacting loose flakes or blocks. During the drilling of bolt holes one should make observations as a check on the basis that decisions are based up on. One can then, in time, make alterations in the program if found necessary. It is also important that the person undertaking this type of work should have a general knowledge of the construction, the district's rocks, topography and zones of weakness and also the best possible experience in rock support. On some occasions the problems are simple and obvious, but one should also be aware that in general rock support is a very difficult problem complex which requires experience and where erroneous conclusions can have serious economic and dangerous consequences. It is therefore recommended that experts should be consulted as far as is possible.

To build up a picture of the background to a stability problem requires an intimate knowledge of the contributing factors, how they manifest themselves and can be observed and in what degree each of the factors affect the specific application being considered. To discuss all the factors for such a study is outside the scope of this handbook. It must therefore be clearly stated that the following division and discussion of stability problems are considerably simplified and incomplete. The treatment is built upon the simplest observations which can be made in a rock cavity, and the simple conclusions that can be drawn from these without entering into details or variations.

As a main division of stability problems where bolting is under consideration we have selected the following five groups:

A. Foliated and bedded rocks
B. Blocky rock
C. Broken rock
D. Clay filled fractures and crushing zones
E. Rock pressure problems.

These groups are handled separately in order to simplify the problems so as to give a pointer to the methods of attack. Over-lapping will inevitably occur since most stability problems are a combination of one or more of the group plus other factors, e.g. water conditions. In principle, there are four variable main factors which are the determining ones for stability problems.

1. Rock pressure - the stresses in the rock at any time due to gravitational effects, earth movements, cooling stresses, blasting of the excavation, etc.

2. Material properties - the properties of the rock in its widest meaning, i.e. particle size, cementing material, compressive and tensile strength, elasticity, mechanical anisotropy, etc. The latter term includes fracture and fault frequency and directions, roughness and possible fracture infilling. Fracture and crushing zone properties.

3. Water conditions - pore water pressure and the pressure and flow lines modification by the blasting. Drying out or wetting of the rock, larger water inflows, washing out of fracture infillings, etc.

4. Geometric conditions - here is meant the interaction between the orientation of the cavity and all the directional effects which influence stability. As examples can be cited, the importance of the angle at which a clay filled zone crosses a cavity or the orientation of the principle stresses to the contour of the excavation in spalling ground.

When we now progress to consider the separate types of conditions, A to E as a defined earlier, we must be aware that there are special cases of the variations in the main factors 1-4. We should therefore bear this in mind and consider how these are altered with time in each case.

A. FOLIATED AND BEDDED ROCK

Such rock is found when the material properties are such that there is a dominant parallel oriented planar system in the rock. This can occur in many ways, e.g. in a gneiss with dark and light bands, where dark mica occurring on defined planes can give sliding surfaces or cleavage surface. Such cleaving is also found in slates chlorite, shales, etc.

Other typical rocks of this type are phyllites and mica schists. Rocks which cannot be defined as slatey but which have a dominant plane cleavage system can be treated in the same way for stability considerations.

The normal types of bolts are usually especially suitable for such conditions. The stability problems arise due to the rock properties, as outlined above, combined with a geometric consideration showing that the weak planes are oriented parallel or nearly parallel with the contour of the cavity, see fig. 49 and 57. Factors such as the width of the cavity, how easily the rock cleaves, the frequency of the planes

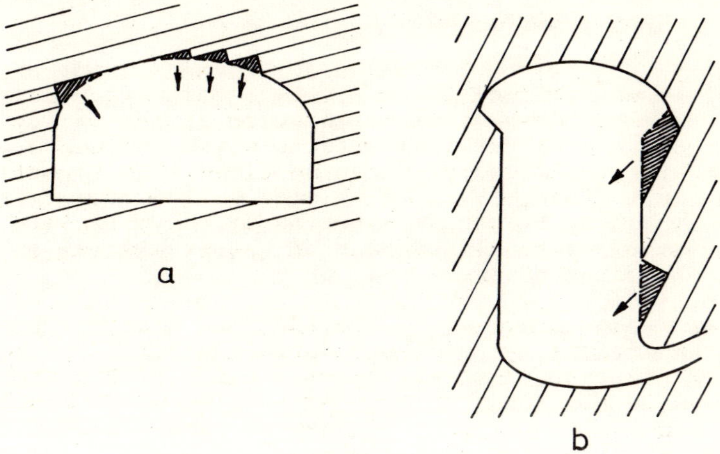

Fig. 57. Rock fall in absence of support

of weakness, the elasticity of the rock, etc. will be of importance. If we ignore rock pressure, which is dealt with separately, support of such rock will be by holding together a number of layers such that it does not flake and disintegrate. Very often it is advantageous to use pretensioned bolts such that the friction in the fractures can be retained or even increased.

In this instance all the bolt types are satisfactory and it will be the practical considerations which determine the types and method of working adopted e.g. expansion bolts at the face, Bergjet bolts for permanent support. The bolt length and the bolting pattern can be selected from the formulae or rules of thumb discussed earlier but it is often permissible to use both shorter bolts and greater bolt distances than the calculated values if the problem is solely of this type.

The elasticity of the individual rock slabs, possible additional fracture systems or weaknesses are the factors which determine whether it is possible to reduce bolt length and/or increase bolt distance.

B. BLOCKY ROCK

In the chapter on determination of bolt lengths and bolt patterns we started by considering single limited loose blocks. When we now consider blocky rock we have a much more difficult situation as we can no longer consider each block separately. We can also say that the material properties have changed from the foliated and bedded rock handled in the previous paragraphs in that we now have several distinct slip planes. These slip planes which give the rock its blocky structure and unstable character can number between one and four fracture planes. They can be planes of foliation, they can be narrow cracks filled with clay minerals, or slip planes with chlorite or graphite coatings. All these types can be combined in a multitude of ways. Although we are deliberately ignoring rock pressures at this stage it should be stated that stability problems in blocky rock very often occur at low rock pressures. The reason for instability is that the individual block is insufficiently restrained and has therefore small forces normal to the boundary faces. A single block can fall out and release more blocks. Such problems are normally met in cavities at shallow depths e.g. in outcrop zones.

We have decided to classify rock as blocky when the variation in block size is not too great and where the thickness of individual slip planes of whatever type, is less than 2-3 cm. The distance between the main fracture planes is over 1 meter. Slip planes wider than 2-3 cm are dealt with separately, see page 74.

Support of blocky areas is treated in the same manner as in the section on individual loose blocks, see page 53. We estimate the average dimensions of the blocks and select the bolt length to reach 1-2 meters into solid rock. Particularly for very wide cavities it is necessary to compare the selected bolt length with that suggested by the formula.

The bolting pattern is determined on a volume/weight assumption in order to estimate the loading on each bolt. It should also be decided whether it is the loading system or the fact that each block requires at least one bolt that will determine the pattern. The strength of the bolt selected will be a determining factor in this. One must also consider the geometrical conditions i.e. the relative orientation of the fractures and the cavity as there are situations where bolting of certain key blocks is sufficient to stabilize the whole and these opportunities should not be over-looked.

At the other extreme we have situations with a well developed steep fracture system in the longitudinal direction of the cavity with a further steep system at right angles and an easily cleaved horizontal foliation. In this condition bolting can be considerable.

C: BROKEN ROCK

If we go a further step in the alteration of rock properties such that the fracture spacing is smaller, i.e. less than 1 meter, we arrive at a type of rock which can be called broken rock. This division is chosen because normal practical bolting patterns only exceptionally give distances less than 1 meter.

The foregoing paragraphs will probably suggest that there should be a bolt in each block. Such a condition can be achieved in blocky rock but with decreasing fracture spacing the limits of accuracy and practicality will soon be reached.

Without entering into details of tests and analyses which have been done in order to determine criteria for bolt spacing, it has become clear that the spacing in most cases can be increased to 3-4 times the average fracture spacing without reducing the factor of safety too much. It could therefore be assumed that the selection of a criterion for classification as broken rock should not be at a fracture spacing of 1 meter but at 0,3m to match with the normal practical minimum bolt spacing of one meter. In practice it is found that the general rule of bolt spacing at 3-4 times the fracture spacing can fail. In the intermediate zone, 1.0 to 0,3m fracture spacing conditions can occur where the friction in the fractures is so low, due to chlorite or clay unfillings, that one must depart from the general rule.

Bolt spacings down to 1 meter can therefore be used for rock with fracture spacings between 0.3 and 1 meter. Often it is necessary to combine the bolts with other means of support when the fracture spacing is small and when friction reducing layers are present

The cause of stability problems in such rock is often the low restraint either alone or in combination with friction reducing layers on one or more of the fracture systems. Also in broken rock the geometrical conditions, i.e. relative orientation of the fracture systems, are of great importance in the degree of difficulty in supporting the excavation while the geometry is not so important in determining the bolting pattern. This is because we nearly always will require straps or wire mesh and the dimensions of these will, to a certain extent, restrict the flexibility of the bolting pattern. The bolt pattern should be such that a reasonable amount of overlapping is allowed and whole rolls of net or full lengths of strapping can be employed. When woven wire mesh is used it is usually best to bolt and secure an area at least as large as the roll of the mesh and then attach the wire mesh with further plates and nuts on each bolt. The attachment plates can well be of a cheaper type.

When bolts are used as support for broken rock it must be done with great care. It is clear that a normal bolting pattern cannot lock more than a small volume around each bolt. There will be broken rock, probably, unaffected by stress even from pretensioned bolts, both between and over the bolts. Bolts alone, or with woven wire mesh must be used therefore only when it is felt that there is no danger of the fall of a larger broken block with one or more bolts in it, but that the bolts will lock off the larger volume and the mesh catches the odd smaler stones which fall. If there is danger that the whole volume round one or several bolts can fall out together with the bolts, than the use of rock straps will reduce the danger. As the straps only seldom can be pulled in to the rock contour, they will not give any active support. A movement will be able to start but the movement will be halted before a larger fall occurs and the movement which has taken place is easily observable and can be repaired. If conditions are particularly difficult bolt straps and wire mesh can be used together, alternatively bolts and shotcrete, either reinforced or unreinforced. These choices must be seen in connection with the permanent support, consideration of the time variable factors and the desired factor of safety.

Recommended bolt lengths for the different spans can be calculated from the formula given earlier, but for use of bolts alone or with only woven wire mesh it is advisable to use larger lengths than the formula suggests.

D CLAY FILLED FRACTURES AND CRUSHING ZONES

We have earlier dealt with situations where the general material properties over an area have been so poor that support was required. Bolting can only be considered for concentrated clay filled fractures or crushing zones under certain clearly defined conditions, and one of these is that the zones must be less than 3 meters wide. This boundary is not as definitive as the above statement would indicate but is stated to make it clear that the limitations of bolting in such conditions are severe. We can also limit ourselves to zones where the neighboring rock is reasonably good and not broken or clay impregnated.

Clay filled fractures and zones have a very variable character and it is impossible to describe them fully but the most important features require a mention. When considering clay zones it is the material properties which vary from case to case. Some zones contain a very high percentage of clay minerals while others can contain fresh rock in sugar lump sized cubes with a 2-3mm clay coating on all surfaces. Other cases occur where 4-5 parallel slips of 10-20 cm width with equally wide bands of relatively intact rock between them. It is therefore important to determine the approximate proportion of clay material present.

It is also important to know what the clay itself is composed of and particularly if it contains swelling minerals. Normally only a small percentage of the clay is of swelling minerals but if there is a considerable percentage of montmorillonite minerals (swelling minerals) present then a lump of relatively dry material taken from such a zone will form a slurry within minutes, or perhaps even seconds when immersed in water. If necessary, it is possible to obtain an indication of the mineral content by means of simple colour tests.

Rock pressure, even if we neglect the high pressures which give rise to spalling, has an effect, in that it can lead to a consolidation of the clay. A well packed material will often be more stable than a loose clay.

The moisture environment has a great influence on the behavior of a clay zone. If the in situ conditions are moist then the clay will have a low shear strength and can easily flow out. If it is dry or is dried then it will normally stabilize. Swelling clays only swell on absorption of water, i.e. a mass which is already wet can contain swelling minerals without this constituting a risk, but the same mass in a dry condition can begin to swell simply due to the increase of air humidity when ventilation is stopped.

Swelling materials in such zones can develop a high pressure if volume increase is prevented. If possible, it can therefore be advantageous to let the mass increase in volume by a few percent before the support takes up the loading.

To summarise these considerations in the simplest way we can say that the most difficult problems occur when the clay zone has:

- a high percentage of clay
- a high percentage of swelling minerals
- a high consolidation
- a dry condition

If we expose such conditions in a cavity then problems will not normally occur immediately. However, if in such a situation we use a stiff support system which does not allow any increase in volume yet allows excess water (e.g. submersion) then the only reliable support is full concrete lining.

The geometrical conditions are also important for clay zones. A steep zone parallel with the longitudinal axis of the cavity, as described before, will always give rise to a very expensive and serious problem.

In the clay itself bolts will not obtain a satisfactory hold. Even cement grouted or polyester grouted bolts, which are the best which could be used, will be so far from reaching a sound anchorage that it is futile even to attempt it. This is the reason that we limited consideration of bolting clay zones to those with good adjacent rock. The bolts are therefore set in the sidewalls. For temporary support, bolts can be used alone for supporting the sidewalls until the permanent support is installed. There will be one or two rows of bolts on each side of the clay zones. One should then expect some fall out of clay from the zone and to prevent this, woven wire mesh can be stretched across. Permanent support of clay zones has been obtained with bolts and reinforced shotcrete, filling or similar methods on many occasions with good results.

Owing to the complicated material situation in clay zones, such methods should be used with great care in doubtful conditions or where a high degree of safety is required.

E ROCK STRESS PROBLEMS

Rock pressure in a rock cavity can manifest itself in many ways and can also have many different causes. In general, rock pressure problems occur when the operating forces, stresses, exceed the loads that the rock is able to sustain. The factors which determine whether rock pressure problems occur are many:

the absolute values of principles stress
the ratio of principle stresses
the rock's strength properties
the variation in strength properties with direction
the geometrical shape of the cavity
the relative orientation of the above factors

In addition, there are the time dependent factors and other practical occurrences such as drying out, water absorption, nearby blasting, temperature fluctuations etc.

The variations in the factors which affect the stability are reflected in the way rock pressure manifests itself. We can observe a range of behavior in the cavity from squeezing of almost plastic material into the opening to sharply cracking spalling where splinters and flakes of rock shoot out of the surface with a sharp report and great force.

It would take too long to review how rock pressures occur, how the risk of rock pressure problems is assessed and how the cavity should be placed and oriented in order to reduce the problems. We therefore assume that these are factors which have already been determined and that we are faced with some manifestation of rock pressure in a tunnel and need to support the tunnel.

In order to understand the principles of support against rock pressure we will examine some simple, typical causes and occurrences. Fig. 58 shows the cross section of a tunnel and the orientation and magnitude of the principle stresses at the section before blasting of the tunnel. For those who do not know the theoretical background for stress analysis is should be stated that all the stresses are compressive. The rock stresses can be measured in several directions at the point at the center of the mass. The largest compressive stress is found along the direction marked σ_1 and the least along the direction σ_2. For the sake of simplicity σ_3 (which is at right angles to the plane of the paper) is assumed to be equal to σ_2. In intermediate directions we will measure a stress σ_1 which lies somewhere between σ and

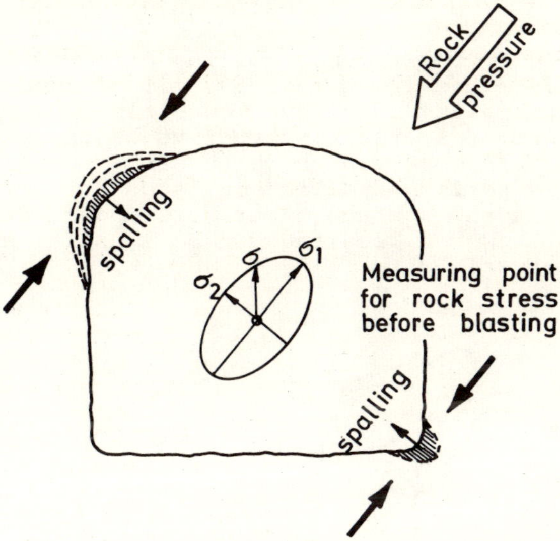

Fig. 58. spalling

σ_2 in magnitude.

For a typical spalling situation to occur σ_1 is usually much larger than σ_2. Also the rock is brittle, stiff, has a low tensile strength and is relatively homogeneous. When the pressures are high enough the spalling will normally occur at the left shoulder where the major principle stress σ_1 is tangential to the rock surface. In serious occurrences the opposite corner will also be affected. When σ_2 and σ_3 are alike, as in this example, the face will also spall if the rock properties are the same in all directions. What happens is that the compressive stresses along the given directions become large enough to induce high tensile stresses in directions at right angles and thin splinters and flakes (discus shaped) shoot into the tunnel. The compressive stresses are, for differing reasons, concentrated at the rock surface where the induced tensile stresses are then concentrated to an even larger extent. For each rock fragment which is released the stresses become concentrated at the newly exposed surface and will release further flakes.

In difficult conditions a spiring effect can be built up in this manner, see fig. 59. It will be seen from this that scaling under such conditions is risky work, and is also of little constructive assistance. If we scale off loose flakes which still remain in place we risk restarting spalling by reconcentrating the stresses at the surface. For more moderate rock stresses, however, scaling can be a useful and often sufficient safeguard.

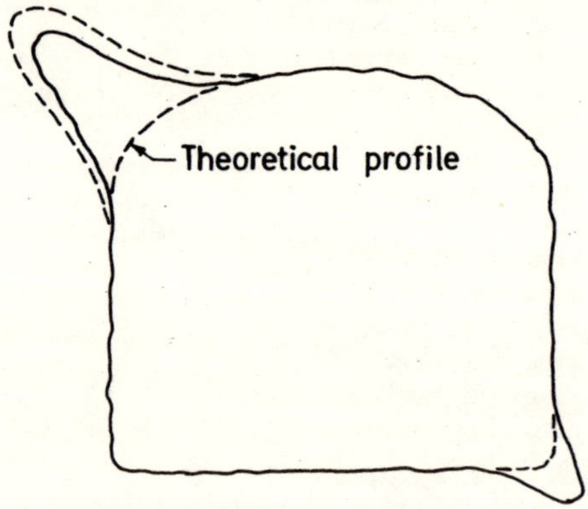

Fig. 59. Unrestrained development of spalling

Against very heavy spalling it has been usual to support right to the face with expansion bolts. These bolts cannot prevent deformation or the scaling off of rock flakes, but they prevent the falling out of the flakes and offer some restraint to the tensile stresses. By this method the stresses are spread into the rock so reducing the surface stresses to a level which the rock is able to bear.

Formulae for dimensioning of bolt support against spalling cannot be given. The stresses can be measured and calculated, but if one was to attempt such a solution it would be too lengthy to assist in a situation where the face must be stabilised for safety reasons. Support of spalling is therefore done by judgement and experience. When a round is fired one usually has a certain "stand-up time" before spalling begins. The insertion of bolts should therefore, if possible, make use of this time. Bolt lengths can be calculated using the formulae as before, but bolts should seldom be shorter than 1.5 meters to obtain a satisfactory spreading of the stresses into the rock. The bolting pattern is most dependent on how one is to prevent fall out between the bolts. Normal bolting patterns can often be used together with straps or wire mesh. On long sections of spalling ground this can take up a large amount of labor time and small bolt distances (down to 0.7 x 0.7 m) with large plates (up to 60 cm sides) have been used to good effect. In this way a satisfactory area is covered.

In order to relieve the rock in cases of intense spalling it can be useful to permit a certain amount of deformation before the bolts take their full load. It has been found that if expansion bolts are fully tensioned they can be over-loaded coursing them to break. Under such conditions it is often sufficient to have the nut hand tight. The rock will then have to deform a little before the bolt becomes loaded and this is usually sufficient to enable them to carry the total load. Under these extreme conditions a deformation before the installation of permanent support is desirable. Steel bolts with a relatively large yield at failure are to be recommended for these conditions. Stiff concrete constructions directly against the rock will often be destroyed if they are installed right at the face.

The extreme spalling situation which has just been described is, in fact, not so risky as it might appear. The problems are both audible and visible and the work can be adopted to this. Rock pressure problems can also occur and often manifest themselves as the development of loose flakes well behind the face or a long time after blasting. This situation can be dangerous when it is often not detected before fall out occurs. Scaling and bolt support in the normal manner is usually the suitable solution. In strongly anisotropic rocks such as shales, phyllites, gneisses etc., it is well to be aware of the conditions which will occur when the foliation direction varies. One can have long stretches of a tunnel where the foliation is such that the tunnel cuts through it. In moderate rock pressure these stretches will probably give no problems. If, however, the orientation changes such that the foliation is tangential to the tunnel surface, then a potentially dangerous situation can occur in that scabbing and fall-out occur after weeks or even months. The reason that such scabbing can be concentrated in such sections is the considerably lower tensile strength of the rock in the direction across the foliation compared with the longitudinal one.

Rock pressures can be too low to give rise to scabbing in the first instance but high enough in the tangential case. In elastic and plastic rocks such as greasy phyllits, etc., such scabbing can take place without audible noise.

All the special situations which can arise from varying rock pressure and rock conditions in tunnels, power station halls and other large cavities cannot be discussed here. For support of stability difficulties which can be accredited to rock pressures it will almost always be necessary to use bolts which are very suitable devices for such problems. It is important to be aware that we can rarely expect the bolts to prevent deformation and they are therefore usually dimensioned on the basis of the load they will carry. It is often a case of mounting sufficient bolts, of a suitable length, which experience has shown to be satisfactory. The bolts ensure that we allow the rock a controlled amount of deformation so relieving the load to values which the bolts can support. Alternatively we can describe it as follows: the forces occurring in the critical surface layers are reduced by means of deformation and the stresses are transferred in the necessary extent to the rock behind the surface.

In conclusion, we would like to remark that the considerations for support by bolts should also include an estimate of the phsychological and personal side of the problem as well as the purely technical. It often occurs that the morale at the working site is poor and this can be due to the uncertain rock conditions. Difficult rock carries with it problems for the tunnel crews in the form of more and heavier work, perhaps a lower remuneration and not least a possible general feeling of insecurity. Such problems will probably first arise when the rock has been difficult for a period or there has possibly been a fall or accident.

Even if such areas are supported, or are to be satisfactorily supported by bolts and possible accessories, one can experience that some scepticism towards bolt support still gives a feeling of dissatisfaction. It can also happen that it becomes difficult or impossible to obtain people to do the bolting work.

One should first and foremost attempt to avoid such conditions by supporting before conditions become serious. One should also be aware that the work does carry a risk and that better equipment may be necessary. We refer here to the most risky part of the work which is the drilling of bolt holes, if this is done with hand held machines. If only hand held machines are available in a difficult situation, the work must start from under a safe roof and must be continued progressively in the order scale-drill-bolt-tension. Then the work can continue with the next bolt and the work advances in steps into the poor area.

Conditions can be so bad that hand held equipment cannot be used. In these cases a hydraulic rig specially built for the purpose is the only alternative. Such rigs are available from several sources and they give a greatly increased safety of working together with increased capacity. In large construction it is therefore desirable to have such equipment available.

It also happens that people are sceptical about the bolts themselves and say, for example, "expansion shells do not grip this rock". In such situations it is helpful to have tensioning equipment which gives a measurable stress which the persons themselves can check. The hydraulic bolt tensioner previously described is an ideal device in this connection.

If a bolting rig and proper tensioning device are used together with the necessary instruction, problems of confidence in the support system will seldom occur. This presupposes that the correct bolt type is being used and that the particular stability problem is suitable for practical solution with rock bolts and accessories.

10. SPECIAL BOLT SYSTEMS

TRUSS BOLTING.

The principle of this system is shown in fig. 60. Two angled bolts are connected by a horizontal rod which has a central turnkey for tensioning such that a compression zone is formed between the bolts. By mounting a number of such sets parallel with each other the zone can be increased in length. The method was developed and patented by the Birmingham Bolt Company in England and is used especially in coal mines. In Norway, as far as we are aware, it has only been used in Dalen Mine at Brevik.

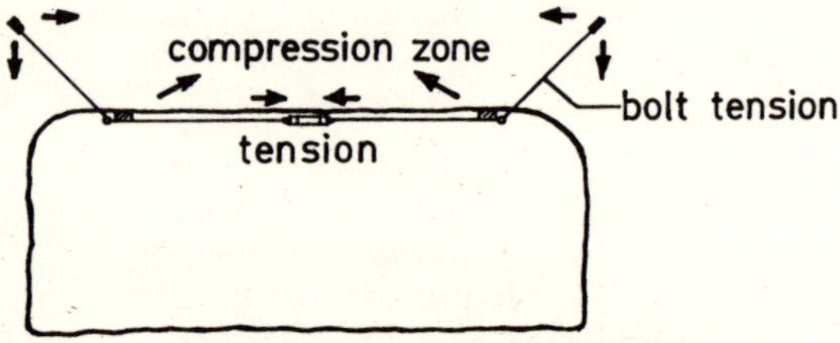

Fig. 60. Truss bolting

GLASS FIBRE BOLTING.

At the Spokane Mining Research center of the U.S.

Bureau of Mines they have developed a new bolt type, fig. 61., where the bolt consists of a bunch of fibre glass threads and the grout of polyester and hardener. The fibreglass "rope" is held on a spool and is fed into the hole at the same time as polyester grout is injected. The holes need not necessarily be straight and the depth can vary as the fibreglass bunch is cut off outside the hole. Tests show that the bond strength between the polyester and the hole wall for a 34 mm diameter hole was 5.2 tons/meter for a wet hole and approx. 37 tons/meter for a dry hole. The material costs for a one meter bolt at present prices is approx. ($1.50), but complicated and expensive equipment for installation will increase the total costs considerably.

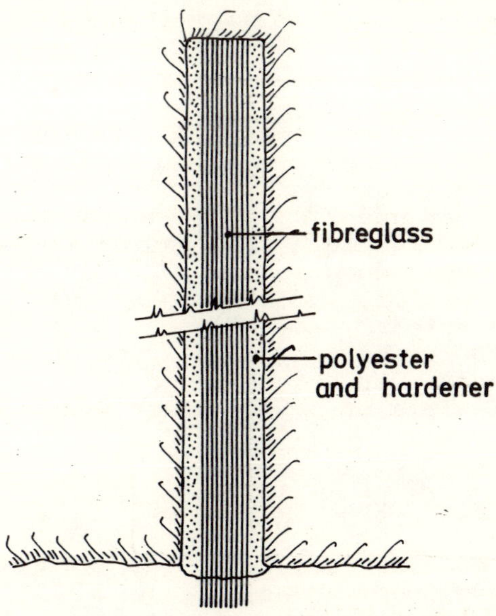

Fig. 61. Pumpable fibreglass bolt

FORE-POLING

The method entails drilling of long boreholes (10-15 m) at a small angle to the drivage axis and grouting in rebar bolts before the blasting of the tunnel, see fig. 62.

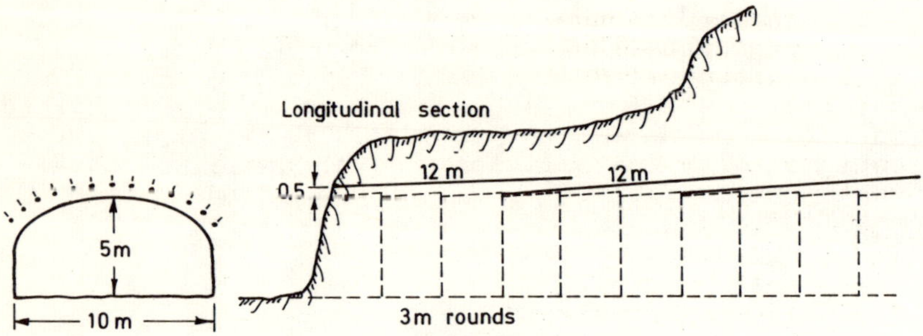

Fig. 62. Forepoling

Forepoling in this way can be used to reduce the length of entrance cuttings in places with little rock cover or in a tunnel section with little or poor rock cover under a lot of loose material. By this means is created a longitudinal reinforcement of the roof between 0,5 and 1,0 meter outside the theoretical profile which is operative when the tunnel is blasted. Depending on the length and dimension of the bolts and the type of grout being used, the bolts can be mounted using either the Bergjet principle or by injection, see fig. 29, page 36.

PLASTIC EXPANSION SHELLS.

A new type of plastic (Delrin 100) expansion shell has been developed in Sweden. The shell consists of 4 blades with two conical sections. As shown in fig. 63, the shell has two conical sections such that its expansion is parallel. The smaller cone has an end stop so that the expansion is limited. The blades have a groove near the top and bottom for a strong rubber ring (O ring). When the bolt has been

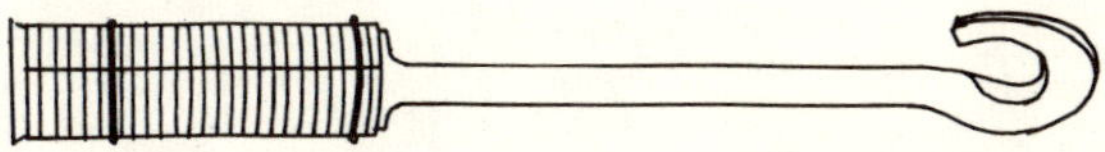

Fig. 63. Hanger bolt with plastic expansion shell

used for example, for hanging a ventilation tube, the bolt is screwed out a little and given a blow. The expansion shell then collapses and the bolt and shell can be removed from the hole for further use. Tests have shown that the plastic shell begins to slip in granite at a load of 5.5 tons.

Resin-grouted Rockbolts of Fibreglass, fig. 64.

A recent developed bolt of fibreglass (Celtite, Moon) has been used as working support in some mines where the rock later had to be removed. Ordinary rockbolts of steel have the disadvantage of remaining intact after blasting, and in coal mines they would destroy the cutters. Fibreglass bolts are destroyed by blasting and does not prevent the mucking up. The inner end of the bolt is cut of in an angle of 45° to make the mixing of polyester resin capsules better. The outer end has a wedge part 10 - 15 cm long on which a brick and a socket are being applicated. The wedge is hardwood and is put on the bolt by the manufacturer. The bolt contains 73% fibreglass and the rest polyesterresin. The dimensions are 18, 20 and 22 mm dia. The breaking, shear and bending stresses are 6200, 3000 and 7500 kp/cm^2, respectivly. The breaking load, however, is depended on the bonded length.

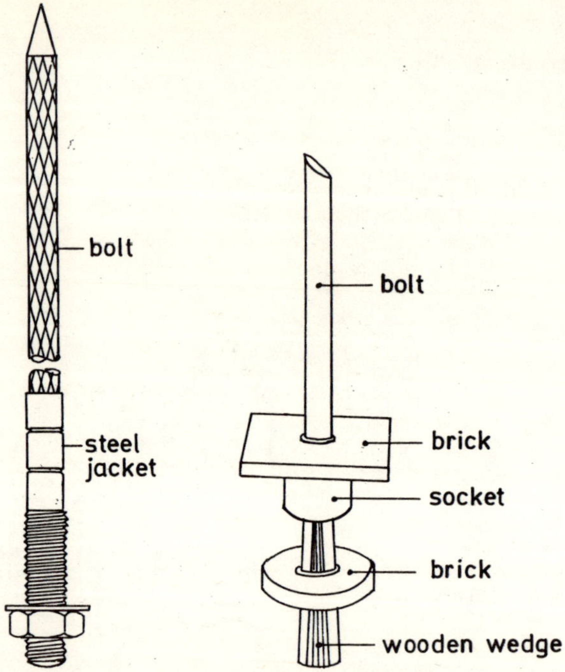

Fig. 64. Resin-grouted rockbolts of fibreglass

The latest rockbolt of fibreglass is made in Austria and supported at the end with a steel jacket with threads, fig. 64, which enables it to be tensioned.

Due to rapid developments in the bolting domain and a constant flow of new types of rockbolts onto the market, it is difficult to keep up-to-date. It is most important, therefore, to follow the manufacturer's instructions and procedures for the actual kind of bolt being used.

When the work is completed, it must always be in accordance with the required safety regulations and fully satisfy the demands of public safety.